Microcognition

Explorations in Cognitive Science
Margaret A. Boden, general editor

1. *Mental Processes: Studies in Cognitive Science,* H. Christopher Longuet-Higgins

2. *Psychosemantics: The Problem of Meaning in the Philosophy of Mind,* Jerry A. Fodor

3. *Consciousness and the Computational Mind,* Ray Jackendoff

4. *Artificial Intelligence in Psychology: Interdisciplinary Essays,* Margaret A. Boden

5. *Computers and Thought: A Practical Introduction to Artificial Intelligence,* Mike Sharples, David Hogg, Chris Hutchison, Steve Torrance, and David Young

6. *Microcognition: Philosophy, Cognitive Science, and Parallel Distributed Processing,* Andy Clark

Microcognition:

Philosophy, Cognitive Science, and Parallel Distributed Processing

Andy Clark

A Bradford Book
The MIT Press
Cambridge, Massachusetts
London, England

Second printing, 1990

This book was set in Palatino by Asco Trade Typesetting Ltd., Hong Kong, and printed and bound in the United States of America.

Library of Congress Cataloging-in-Publication Data

Clark, Andy.
Microcognition: philosophy, cognitive science, and parallel distributed processing.

(Explorations in cognitive science; [6])
1. Connectionism. 2. Human information processing. 3. Artificial intelligence.
4. Cognitive science. I. Title. II. Series: Explorations in cognitive science; 6.
BF311.C54 1989 153 88-37282
ISBN 0-262-03148-5

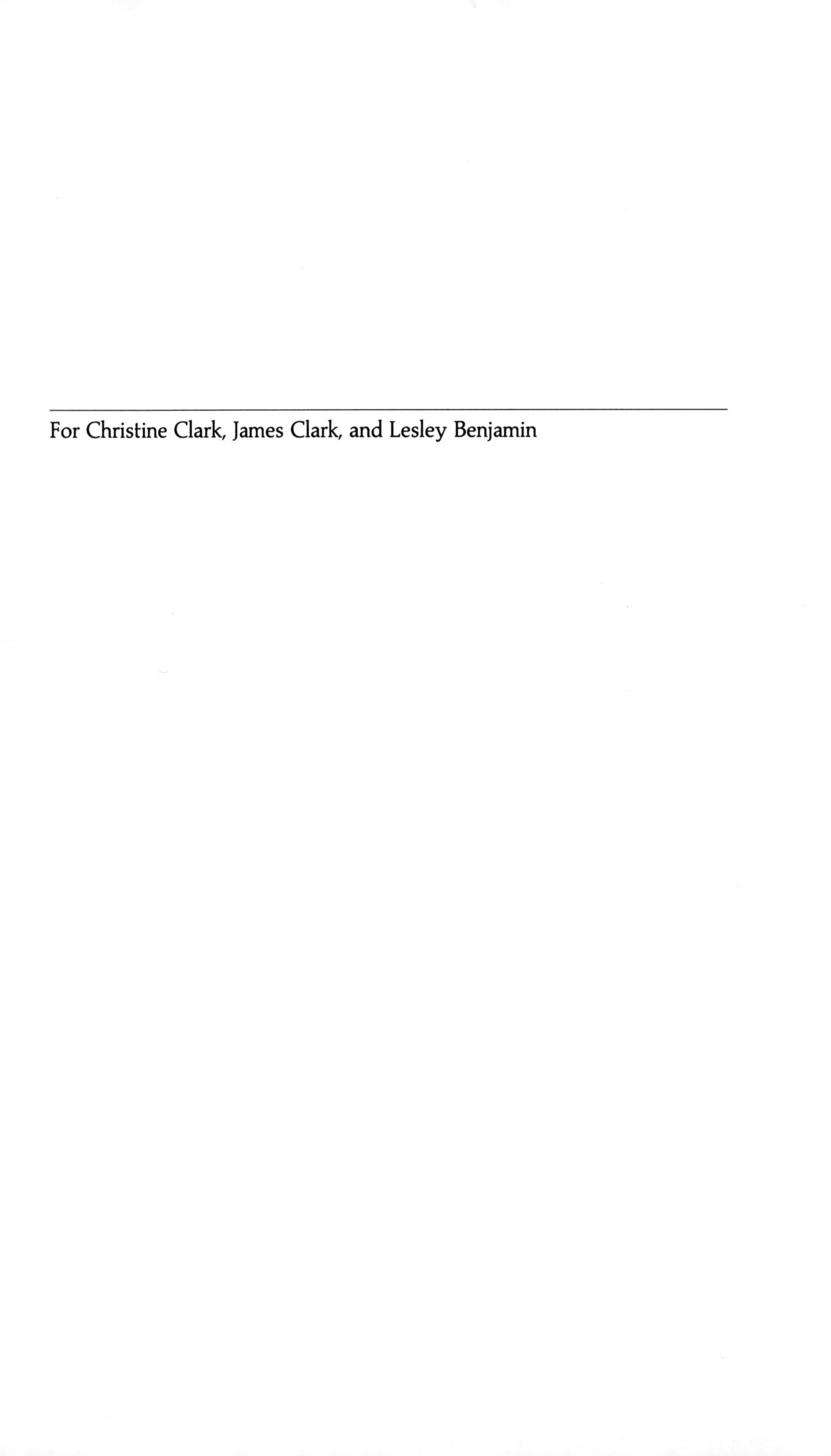

For Christine Clark, James Clark, and Lesley Benjamin

Contents

Preface

The subtitle is enough to put anyone off: *Philosophy, Cognitive Science, and Parallel Distributed Processing*. It might have read: the elusive, the ill-defined, and the uncharted. For all that, the project thrust itself upon me with an unusual sense of its own urgency. Parallel distributed processing (PDP) is an exciting and provocative new movement within cognitive science. It offers nothing *less* than a new kind of computational model of mind. The bad news is that it as yet offers nothing *more* than a hint of the nature and power of such models. But the hints themselves are remarkable and have the potential, I believe, to reshape both artificial intelligence (AI) and much of the philosophy of mind. In particular, they offer a new picture of the relation between sentences ascribing thoughts and the in-the-head computational structures subserving intelligent action. The final product of such reshaping will not be found in this work. At best, I offer some views on the central points of contrast between the old shape and the new, a personal view on most of the major issues, and along the way, a reasonably detailed taxonomy of features, distinctions, and subprojects. The taxonomy, though necessarily individualistic, may be of some use in future discussions of what is, in effect, a whole new topic for philosophy and AI. The conclusions are often provisional, as befits discussions of an approach that is still in its infancy. By the time this book sees print, there will be many new and relevant developments. I hope the book will provide at least a framework in which to locate them.

The reader should be warned of my peculiar circumstances. I am first of all a philosopher, with only a secondary knowledge of AI and evolutionary biology. I am fortunate to work in the highly interdisciplinary School of Cognitive and Computing Sciences at Sussex University. It is only thanks to that harsh selective environment that I have been able to avoid many glaring misunderstandings. Those that remain are, in the time-honored clause, entirely my own responsibility. Adaptation, even to an environment of AI researchers and cognitive scientists, falls somewhat short of an optimizing process.

Acknowledgments

Special thanks are due to the following people (in no particular order): Neil Tennant and Donald Campbell for showing me what a biological perspective was all about; Martin Davies, Barry Smith, and Michael Morris for reminding me what a philosophical issue might look like; Aaron Sloman for encouraging an ecumenical approach to conventional symbol-processing AI; Margaret Boden for making the School of Cognitive and Computing Sciences possible in the first place; my father, James Henderson Clark, and my mother, Christine Clark, for making me possible in the first place; my father and Karin Merrick for their saintly patience with the typing; H. Stanton and A. Thwaits for help with the many mysterious details of publishing a book; the Cognetics Society at Sussex (especially Adam Sharp and Paul Booth) for some of the graphics; and Lesley Benjamin for invaluable help with the problem of meaning in all its many forms.

I would like to thank the copyright owners and authors for permission to use the following figures. Figures 4.3 and 4.4 are from S. J. Gould and R. Lewontin, "The Spandrels of San Marco and the Panglossian Paradigm," *Proceedings of the Royal Society of London*, Series B, 205 (1979) no. 1161; 582–583. Reproduced by permission of the authors and the Royal Society. Table 5.1 and figures 5.1, 5.3, 5.4, 5.5, and 5.7 are from J. McClelland, D. Rumelhart, and the PDP Research Group, *Parallel Distributed Processing: Explorations in the Microstructure of Cognition*, vols. 1 and 2 (Cambridge: MIT/Press, 1986). Reproduced by permission of the authors and the MIT Press.

Parts of chapters 3 and 4 are based on the following articles of mine that have appeared in scholarly journals. Thanks to the editors for permission to use this material.

"From Folk-Psychology to Cognitive Science," *Cognitive Science* 11, no. 2 (1987): 139–154.

"A Biological Metaphor," *Mind and Language* 1, no. 1 (1986): 45–64.

"The Kludge in the Machine," *Mind and Language* 2, no. 4 (1987): 277–300.

Adapted versions of parts of chapters 7 and 9 are due to appear as:

"PDP or Not PDP?" in S. Torrance and R. Spencer-Smith, eds., *Philosophy and Computation* (Norwood, N.J.: Ablex).

"Connectionism and the Multiplicity of Mind," in *AI Review*, special issue on connectionist models.

Introduction
What the Brain's-Eye View Tells the Mind's-Eye View

1 *After the Goldrush*

Cognitive science is the goldrush of the mind. Everybody's searching for it. Worse still, everybody (symbolic AI workers, subsymbolic AI workers, neuroscientists, naturalistic philosophers, and so forth) claims to be finding it. Or at least, they claim to know where to look.

For all that, I believe the place of mind in cognitive science is highly problematic. In many ways the very idea of mind is intimately tied up with the apparatus of propositional attitude ascription: the use of such sentences as "Mary believes that Goldbach's Conjecture is true" to describe mental states. The use of this kind of apparatus (sentential ascriptions of belief, desire, and the like) is often denigrated as "folk psychology." Its denigrators believe that a good computational account of mental processing will not involve any neat analogues to the concepts and relations visible in daily talk. Its proponents believe that much cognitive activity is mechanistically impossible without such analogues (internal tokens, syntactically identified). The two camps are often also divided in their choice of computational architecture. Those who favor folk psychology (such as Fodor) see conventional symbol-processing AI as the obvious option. Those against folk psychology (such as Churchland) tend to favor connectionism or parallel distributed processing. (If all this is double Dutch, be patient.)

This kind of dispute, I shall argue, is deeply muddled. It is muddled because (1) folk psychology does *not* seek to model computational processes, and its dignity does not depend on there being in-the-head analogues to the propositional attitudes, and (2) parallel distributed processing and conventional approaches to mental modeling need not be uniformly regarded as *competing* paradigms of cognitive architecture: cognitive psychology may require many kinds of computational models for different purposes. This book is largely an attempt to establish these two propositions. Along the way very substantial attention is given to laying out the theoretically interesting differences between conventional AI and connectionist AI (parallel distributed processing). In the remainder of the introduction I shall sketch the main lines that the discussion will follow.

2 *Parallel Distributed Processing and Conventional AI*

Parallel distributed processing names a broad class of AI models. These models depend on networks of richly interconnected processing units that are individually very simple. The network stores data in the subtly orchestrated morass of connectivity. Some units are connected to others by excitatory links, so that the activation of one will increase the likelihood that the other is activated. Some are inhibitorily linked. Some may be neutral. The overall system turns out to be an impressive pattern completer that is capable of being tuned by powerful learning algorithms. Many useful properties seem to come easily with such a setup. Taken together, these allow such systems to represent data in an economical yet highly flexible way. A certain class of work in conventional AI I shall call *semantically transparent*. A model will count as semantically transparent if and only if it involves computational operations on syntactically specified internal states that (1) can be interpreted as standing for the concepts and relations spoken of in natural language (such items as "ball," "cat," "loves," "equals," and so on) and (2) these internal tokens recur whenever the system is in a state properly described by content ascriptions employing those words: the token is, as we shall say, *projectible* to future cases. (Note that such states need not be localizable within the machine. The point is rather that we can make sense of the system as operating according to computational rules on entities of that grain.) In short, a system is semantically transparent if there is a neat mapping between states that are computationally transformed and semantically interpretable bits of sentences. A great deal of work in conventional AI (but not all) is semantically transparent. Work in a highly distributed, connectionist paradigm is not. And therein, I shall argue, lies a philosophically interesting difference.

3 *The Multiplicity of Mind*

Some PDP theorists argue that conventional AI models are at best good approximations to the deep truth revealed by connectionism. Some conventional theorists argue that connectionism displays at best a new way of implementing the insights contained in more traditional models. Both camps are thus endorsing what I shall call the uniformity assumption. This states that a single relation will obtain between connectionist and conventional models for every class and aspect of mentality studied by cognitive psychology.

The uniformity assumption is, I believe, distortive and unhelpful along a number of dimensions. Most straightforwardly, it is distortive if (as seems likely) the mind is best understood in terms of a multiplicity of virtual machines, some of which are adapted to symbol processing tasks and some of which are adapted for subsymbolic processing. For many tasks, our everyday performance may involve the cooperative activity of a variety of such machines. Many connectionists are now sympathetic to such a vision. Thus, Smolensky (1988) introduces a virtual machine that he calls the conscious rule interpreter. This is, in effect, a PDP system arranged to simulate the computational activity of a computer running a conventional (and semantically transparent) program.

Less straightforward but perhaps equally important is what might be termed the multiplicity of *explanation.* This will be a little harder to tease out in summary form. The general idea is that even in cases where the underlying computational form is genuinely connectionist, there will remain a need for higher levels of analysis of such systems. We will need to display, for example, what it is that a number of *different* connectionist networks, all of which have learned to solve a certain class of problems, have in common. Finding and exhibiting the commonalities that underpin important psychological generalizations is, in a sense, the whole point of doing cognitive science. And it may be that in order to exhibit such commonalities we shall need to advert to the kinds of analysis found in symbolic (nonconnectionist) AI.

4 *The Mind's-Eye View and the Brain's-Eye View*

AI that involves conventional, semantically transparent symbol processing is to be identified with what I am calling the mind's-eye view. The mind's-eye view generates models based on our intuitive ideas about the kind of semantic object over which computational operations need to be defined. They survey the nature of human thought from within normal human experience and set out to model its striking features. The models produced depend on encoding and manipulating translations of the symbol strings of ordinary language. For some explanatory projects, I shall argue, such an approach may indeed be both correct and necessary. But for others it looks severely limited.

The mind's-eye approach was prominent in the late sixties and throughout the seventies. It is characterized by the tasks it selects for study and the forms of the computational approach it favors. The tasks are what I shall term "recent achievements." "Recent" here has both an evolutionary and a developmental sense. In essence, the tasks focused on are those we intuitively consider to be striking and interesting cognitive achievements. These include chess playing (and game playing in general), story understanding, conscious planning and problem solving, cryparithmetic puzzles, and scientific creativity. Striking achievements indeed. And programs were devised that did quite well at such individual tasks. Chess-playing programs performed at close to the world-class level; a scientific creativity program was able to rediscover one of Kepler's laws and Ohm's law; planning programs learned to mix and match old successful strategies to meet new demands; new data structures enabled a computer to answer questions about the unstated implications of stories; cryparithmetic programs could far outpace you or me. But something seemed to be missing. The programmed computers lacked the smell of anything like real intelligence. They were rigid and brittle, capable of doing only a few tasks interestingly well.

This approach (which was not universal) may have erred by placing too much faith in the mind's own view of the mind. The entities that found their way into such models would be a fairly direct translation of our standard expressions of belief and desire ultimately into some appropriate machine code. But why suppose that the mind's natural way of understanding its own and others mental states by using sentential attributions of beliefs, desires, fears, and so on should provide a powerful model on which to base a scientific theory of mind? Why suppose, that is, that the computational substrate of most thought bears some formal resemblance to ordinary talk of the mind? Such talk is an evolutionarily recent development, geared no doubt to smoothing our daily social interactions. There is no obvious pressure here for an accurate account of the computational structure underlying the behavior that such talk so adequately describes and (in a very real sense) explains.

Part 1 of the book examines the mind's-eye approach, associating it with a commitment to semantically transparent programming. It looks at some standard philosophical criticisms of the approach (mistakenly identified by many philosophers with an AI approach in general) and also raises some further worries of an evolutionary and biological nature. In part 2 our attention is focused on the PDP alternative, which I call the "brain's-eye view." The label refers to the brainlike structure of connectionist architectures. Such architectures are neurally inspired. The neural networks found in slugs, hamsters, monkeys, and humans are likewise vast parallel networks of richly interconnected, but relatively slow and simple, processors. The relative slowness of the individual processors is offset by having them work in a cooperative parallelism on the task at hand. A standard analogy is with the way a film of soap settles when stretched across a loop (like the well-known children's toy). Each soap molecule is affected only by its immediate neighbors. At the edges their position is determined by the loop (the system input). The affect of this input is propagated by a spreading series of local interactions until a global order is achieved. The soap film settles into a stable configuration across the loop. In a PDP system at this point we say the network has *relaxed* into a solution to the global problem. In computing, this kind of cooperative parallelism proves most useful when the task involves the simultaneous satisfaction of a large number of small, or "soft," constraints. In such cases (biological vision and sensorimotor control are prime examples) the use of a parallel cooperative architecture can make tractable tasks that (with such slow processors) we could not otherwise complete in the time available.

Such approaches are most obviously useful for such evolutionarily basic tasks as vision and sensorimotor control. Such computational achievements are by no means as intuitively central to cognition as, say, chess playing. Nonetheless, hamsters, slugs, and tortoises all sense and move, and they depend on parallel neural networks to enable them to do so. The secret of real intelligence may be revealed by the lessons such humble creatures are able to teach us. For it may be that the flexibility and common sense associated with human cognitive achievements is the result of an underlying computational form chosen by natural selection precisely for its ability to help solve such evolutionarily basic problems.

In examining this conjecture, we cannot afford to ignore the obvious differences between human cognition and, say, the cognitive skills of a hamster. Humans do perform feats of complex, logical reasoning. How is it done? In the closing chapters of the book I examine the conjecture that such reasoning depends on our coming to simulate the serial, manipulative capacities of a more conventional computer.

5 *The Fate of the Folk*

And where does all that leave folk psychology? The position I adopt explicitly rejects what I call the syntactic challenge. The syntactic challenge demands that if beliefs and desires are real and cause behavior, there must be neat, in-the-head syntactic analogues to the semantic expressions in sentences ascribing them. I in general deny this to be the case. Instead, I see belief and desire talk to be a holistic net thrown across a body of the behavior of an embodied being acting in the world. The net makes sense of the behavior by giving beliefs and desires as causes of actions. But this in no way depends on there being computational brain operations targeted on syntactic items having the semantics of the words used in the sentences ascribing the beliefs. Semantically transparent AI posits a neat reductionist mapping between thoughts so ascribed and computational brain states. That is, thoughts (as ascribed using propositional-attitude talk) map onto computational operations on syntactic strings whose parts have the semantics of the parts of the sentences used to express the attitudes. The picture I propose looks like this: thoughts (as ascribed using propositional attitudes talk) are holistically ascribed on the basis of bodies of behavior. Individual items of behavior are caused by computational brain operations on syntactic items, which may not (and typically will not) be semantically transparent. In my model a thought is typically not identical with any computational brain operation on syntactically identified entities, although, of course, there is a systematic relation between brain events and behaviors usefully carved up by ascriptions of beliefs and desires.

6 *Threads to Follow*

Here are some threads to follow if your interest lies in a particular topic among those just mentioned. Conventional, semantically transparent AI is treated in chapter 1, sections 2 to 5; chapter 2, section 4; chapter 4, section 5; chapter 7, section 6; chapter 8, sections 2 and 4 to 8; chapter 9, sections 3, 5, and 6; chapter 10, section 4, and the epilogue. Parallel distributed processing is covered in chapter 5, sections 1 to 7; chapter 6, sections 1 to 8; chapter 7, sections 1 to 7; chapter 9, sections 1 to 7; and chapter 10, sections 2 to 5. Mixed models (PDP and simulated conventional systems) are taken up in chapter 7, sections 1 to 7; chapter 9, sections 1 to 7 (especially 9.6); and chapter 10, section 4. Folk psychology and thought are discussed in chapter 3, sections 1 to 9; chapter 4, section 5; chapter 7, section 6; chapter 8, sections 1 to 9; and chapter 10, section 4. Biology, evolutionary theory, and computational models are discussed in chapter 3, section 6; chapter 4, sections 1 to 6; and chapter 5, section 6.

The main PDP models used for discussion and criticism are the Jets and the Sharks (chap. 5, sec. 3), emergent schemata (chap. 5, sec. 4), memory (chap. 5, sec. 5), sentence processing (chap. 6, secs. 2 and 3), past tense acquisition (chap. 9, secs. 2 and 3).

I

The Mind's-Eye View

The Hitch Hiker's Guide to the Galaxy, in a moment of reasoned lucidity which is almost unique among its current tally of 5, 975, 509 pages, says of the Sirius Cybernetics Corporation products that "it is very easy to be blinded to the essential uselessness of them by the sense of achievement you get from getting them to work at all. In other words—and this is the rock solid principle on which the whole of the Corporation's Galaxy-wide success is founded—their fundamental design flaws are completely hidden by their superficial design flaws."

—D. Adams, *So Long, and Thanks for All the Fish*

Current systems, even the best ones, often resemble a house of cards. The researchers are interested in the higher levels, and try to build up the minimum of supporting props at the lower levels.... The result is an extremely fragile structure, which may reach impressive heights, but collapses if swayed in the slightest from the specific domain for which it was built.

—Bobrow and Winograd, "An Overview of KRL, a Knowledge Representation Language"

Chapter 1
Classical Cognitivism

1 *Cognitivism, Life, and Pasta*

Cognitivism, like life and pasta, comes in a bewildering variety of forms. Philosophers, psychologists, and AI researchers all use the term. For some it is a term of abuse; for others, one of endearment. Like many pseudo-technical terms, its chameleon properties are often used as a convenient antidote to criticism. I therefore propose to use instead the term "classical cognitivism" (and later "conventional AI"). These terms are to signify a conception of mind and computational modeling associated with Newell and Simon's (1976) hypothesis of a physical symbol system and, more generally, with what I call semantically transparent systems, more on both of which below. The point of gracing these particular approaches with the name "classical cognitivism" is simply that in a reasonably precise way it captures the view of the relation between mind and computational modeling that has strongly (indeed, almost exclusively) informed philosophical reactions—both pro and con—to the emerging discipline of cognitive science.[1] This is despite its far from universal acceptance in the AI and cognitive science communities.[2] One goal of this book is to chart some of the limitations of classical cognitivism but to do so without arguing for its total bankruptcy. The justification of such a judgment is, however, a fairly long and involved project that requires a treatment of alternative styles of computational theorizing, a discussion of the role of so-called virtual machines, and a separation of the requirements of psychological explanation from those of the implementation of mindlike qualities in a computer. The present chapter, then, seeks only to outline the classical cognitivist stance, to sketch an associated methodology, and to indulge in a little innocent name dropping.

2 *Turing, McCarthy, Newell, and Simon*

The bigger the names, the harder they drop. These would dent the kinds of floors that supported ancient mainframes. It would be fair to say that Turing made AI conceivable, and McCarthy (along with Minsky, Newell, and Simon) made it possible. Despite occasional pronouncements to the contrary, I think we are still waiting to see it made actual, but more on that in due course.

Turing's (1937) achievement was to formalize the notion of computation itself, using the theoretical device we call a Turing machine. He thereby paved the way for mathematical investigations of computability. But significantly, Turing's formalization also (1) encompassed a whole *class* of mechanisms grouped together not by details of actual physical composition but by their formal properties of symbol manipulation, (2) showed how such mechanisms could tackle any sufficiently well specified problem that would normally require human intelligence to solve, and (3) showed how to define a special kind of Turing machine (the universal Turing machine), which could imitate any other Turing machine and thus perform any cognitive task that any other Turing machine could perform. I shall not review the details of Turing's demonstrations here.[3] For present purposes what matters is that Turing's ideas suggested the notion of machines that, by their formal structure, imitate (and even emulate) the mind. The material stuff (valves, silicon, or whatever) did not matter; the formal properties guaranteed in principle a capacity to perform any sufficiently well specified cognitive task. In the words of a major figure in contemporary cognitive science,

> Turing's work can be seen as the first study of cognitive activity fully abstracted in principle from both biological and phenomenological foundations.... It represents the emergence of a new level of analysis, independent of physics yet mechanistic in spirit. It makes possible a science of structure and function divorced from material substance.... Because it speaks the language of mental structures and internal processes, it can answer questions traditionally posed by psychologists. (Pylyshyn, 1986, 68)

Classical cognitivism, thanks to the work of Turing, was then on the cards. It was some time, however, before classical cognitivism could develop into a viable, experimental discipline. That development required first the arrival of the general-purpose digital computer and second the availability of a powerful and flexible high-level programming language. John von Neumann provided the practical design, and John McCarthy, around 1960, provided a language. The language was called LISP, which stood for list processing, and it made possible the first sustained run of research and development within the classical-cognitivist paradigm.[4] This run of research and development became theoretically self-conscious and articulate with A. Newell and H. Simon's abstraction of the notion of a physical symbol system.

3 *The Physical-Symbol-System Hypothesis*

A *physical symbol system,* according to Newell and Simon (1976, 40–42), is any member of a general class of physically realizable systems meeting the following conditions:

> (1) It contains a set of symbols, which are physical patterns that can be strung together to yield a structure (or expression).
>
> (2) It contains a multitude of such symbol structures and a set of processes that operate on them (creating, modifying, reproducing, and destroying them according to instructions, themselves coded as symbol structures).
>
> (3) It is located in a wider world of real objects and may be related to that world by designation (in which the behavior of the system affects or is otherwise consistently related to the behavior or state of the object) or interpretation (in which expressions in the system designate a process, and when the expression occurs, the system is able to carry out the process).

In effect, a physical symbol system is any system in which suitably manipulable tokens can be assigned arbitrary meanings and, by means of careful programming, can be relied on to behave in ways consistent (to some specified degree) with this projected semantic content. Any general-purpose computer constitutes such a system. What, though, is the relation between such systems and the phenomena of mind (hoping, fearing, knowing, believing, planning, seeing, recognising, and so on)? Newell and Simon are commendably explicit once again. Such an ability to manipulate symbols, they suggest, is the scientific essence of thought and intelligence, much as H_2O is the scientific essence of water. According to the physical-symbol-system hypothesis, "the necessary and sufficient condition for a physical system to exhibit general intelligent action is that it be a physical symbol system," Newell and Simon thus claim that any generally intelligent physical system will be a physical symbol system (the necessity claim) and that any physical symbol system "can be organised further to exhibit general intelligent action" (the sufficiency claim). And general, intelligent action, on Newell and Simon's gloss, implies "the same scope of intelligence seen in human action."

It is important to be as clear as possible about the precise nature of Newell and Simon's claim. As they themselves point out (1976, 42), there is a weak (and incorrect) reading of their ideas that asserts simply that a physical symbol system is (or can be) a universal machine capable of any well-specified computation, that the essence of intelligence lies in computation, and that intelligence could therefore be realized by a universal machine (and hence by a physical symbol system). The trouble with this reading is that by leaving the nature of the computations involved so unspecified, it asserts rather too little to be of immediate psychological interest. Newell and Simon rather intend the physical-symbol-system hypothesis as "a specific *architectural* assertion about the nature of intelligent systems" (1976, 42, my emphasis). It is fair, if a little blunt, to render this specific architectural assertion as follows.

> *The strong-physical-symbol-system (SPSS) hypothesis.* A virtual machine engaging in the von Neumann–style manipulation of standard symbolic atoms has the direct and necessary and sufficient means for general intelligent action.

It will be necessary to say a little about the terms of this hypothesis and then to justify its ascription to Newell and Simon.

About the terms, I note the following. A virtual machine is a "machine" that owes its existence solely to a program that runs (perhaps with other intervening stages) on a real, physical machine and causes it to imitate the usually more complex machine to which we address our instructions (see, for example Sloman 1984). Such high-level programming languages as LISP, PROLOG, and POP11 thus define virtual machines. And a universal Turing machine, when it simulates a special-purpose Turing machine, may be treated as a virtual version of the special-purpose machine.

"Von Neumann–style manipulation" is meant to suggest the use of certain basic manipulatory operations easily provided in a Von Neumann machine running a high-level language like LISP. Such operations would include assigning symbols, binding variables, copying, reading and amending symbol strings, basic syntactic, pattern-matching operations (more on which later), and so on. Connectionist processing, as we shall see, involves a radically different repertoire of primitive operations.

The next phrase to consider is "standard symbolic atoms." This highlights what *kinds* of entities the SPSS approach defines its computational operations to apply to. They are to apply to symbolic expressions whose parts (atoms) are capable of being given an exact semantic interpretation in terms of the concepts and relations familiar to us in daily, or at any rate public, language. These are words (atoms) such as "table," "ball," "loves," "orbit," "electron," and so forth. Some styles of connectionism and many more conventional models (e.g., those of computational linguistics) involve a radical departure from the use of standard symbolic atoms. Since this contrast looms quite large in what follows, it will be expanded upon in section 5 below.

Finally, the locution "*direct* and necessary and sufficient means for general intelligent action" is intended to capture a claim of architectural sufficiency. In effect, the claim is that a strong physical symbol system, as just defined, will be capable of genuine intelligent action. That is, such a machine could be truly intelligent quite independently of any particular underlying architectures (any other real or virtual machines on which it is built), and conversely, it could be so without simulating any other architectures or machines. The SPSS hypothesis thus makes a highly specific and laudably Popperian claim.

What evidence is there to associate such a claim with Newell and Simon? Quite a lot. Some of the evidence comes in the form of reasonably explicit assertions. Some can be inferred from the details of their actual work in AI. And some (for what it is worth) can be found in the opinions of other commentators and critics. A brief review of some of this evidence follows. It is perhaps worth noting that even if Newell and Simon were to deny any commitment to the SPSS hypothesis, the formulation would still serve our purposes, since something like that hypothesis informs the philosophers' view of artificial intelligence (see chapter 2) and, without doubt, still informs (perhaps unconsciously) a great deal of work within AI itself.

4 *Bringing home the BACON*

Still, a little evidence never goes amiss. For a start, we find the following comment sandwiched between Newell and Simon's outline of the nature of a physical symbol system and their explicit statement of the hypothesis: "The type of system we have just defined ... bears a strong family resemblance to all general purpose computers. If a symbol-manipulation language, such as LISP, is taken as defining a machine, then the kinship becomes truly brotherly" (Newell and Simon 1976, 41). Douglas Hofstadter (1985, 646, 664), who takes issue with the idea that baroque manipulations of standard LISP atoms could constitute the essence of intelligence and thought, is happy to ascribe just that view to Newell and Simon.

Moreover, Newell and Simon's own practice does seem to bear such an ascription out. Thus, all their work, from the early General Problem Solver (1963) to their more recent work on production systems[5] and on automating scientific creativity, has been guided by the notion of *serial heuristic search* based on protocols, notebook records, and observation of human subjects. (Heuristic search is a means of avoiding the expensive and often practically impossible systematic search of an entire problem space by using rules of thumb to lead you quickly to the area in which with a little luck the solution is to be found.) For our purposes, the things that most significantly characterize this work (and much other work in contemporary AI besides—see, for example, the AM program mentioned below) are its reliance on a serial application of rules or heuristics, the rather high-level, consciously introspectible grain of most of the heuristics involved, and the nature of the chosen-task domains. I shall try to make these points clearer by looking at the example of BACON and some of its successors, a series of programs that aim to simulate and explain the process of scientific discovery (Langley 1979; Simon 1979; Simon 1987; Langley, et al. 1987).

BACON sets out to induce scientific laws from bodies of data. It takes observations of the values of variables and searches for functions relating the values of the different variables. Along the way it may introduce new variables standing for the ratio of the value of the original variables. When it finds an invariant, a constant relation of the values of different variables, it has (in some sense) discovered a scientific law. Thus, by following simple heuristics of the kind a person might use to seek relations among the data ("try the simple relations first," "treat nonconstant products of ratios between variables as new variables," etc.), BACON was able to generate from Kepler's data "ratios of [successive] powers of the radii of the planets' orbits to [successive] powers of their periods of revolution, arriving at the invariant D^3/P^2 (Kepler's third law), after a search of a small number of possibilities," (Simon 1979, 1088). Similarly, BACON arrived at Ohms law by noticing that the product of electrical current and resistance is a constant.

Now for a few comments on BACON (the point of some of these won't be clear until subsequent chapters, but patience is a virtue). First, BACON makes its "discoveries" by working on data presented in notational formats (e.g., measures of resistance, periods of planetary revolution) that represent the fruits of centuries of human labor. Manipulating these representations could be the tip of the iceberg; creating them and understanding them may constitute the unseen bulk. I say a little more about this in chapters 6 and 7. For now, simply note that BACON and other programs like AM and EURISKO[6] and MYCIN (below) help themselves to our high-level representational formalism. In a recent work Langley et al. (1987, 326) are sensitive to the problem of creating new representational formalisms. But they insist that such problems can be tackled within the architectural paradigm associated with the SPSS hypothesis.

Second and relatedly, the knowledge and heuristics that BACON deploys are coded rather directly from the level of thought at which we consciously introspect about our own thinking. This is evident from Simon's (1987) statement that he relies heavily on human protocols, laboratory notebooks, etc. BACON thus simulates, in effect, the way we reason when we are conscious of trying to solve a problem and it uses kinds of heuristics that with some effort we might explicitly formulate and use as actual, practical rules of thumb. In chapters 5 to 9 I shall conjecture that this kind of thought is a recent overlay on more primitive, instantaneous processes and that though modeling such thought may constitute a psychological theory of such conscious reasoning, it could not serve on its own to instantiate any understanding whatsoever. This level of modeling is common to much but not all contemporary work in AI, including work in expert systems and qualitative reasoning (see, e.g., the section "Reasoning about the Physical World" in Hallam and Mellish 1987). Thus the MYCIN rule (Shortliffe 1976) for blood injections reads: If (1) the site of the culture is blood, (2) the gram stain of the organism is gramneg, (3) the morphology of the organism rod, and (4) the patient is a compromised host, then there is suggestive evidence that the identity of the organism is pseudomonas-aeruginosa (from Feigenbaum 1977, 1014).

Likewise, BACON's representation of data was at the level of attribute-value pairs, with numerical values for the attributes. The general character of the modeling is even more apparent in programs for qualitative-law discovery: GLAUBER and STAHL. GLAUBER applies heuristic rules to data expressed at the level of predicate-argument notation, e.g., "reacts [inputs (HCl, NH_3) outputs (NH_4, Cl)]," and STAHL deploys such heuristics as "identify components: If *a* is composed of *b* and *c*, and *a* is composed of *b* and *d*, and neither *c* contains *d* nor *d* contains *c*, then identify *c* with *d*."

Third, BACON uses fairly slow serial search, applying its heuristics one at a time and assessing the results. Insofar as BACON relies on productions, there is an element of parallelism in the search for the currently applicable rule. But only one production fires at a time, and this is the seriality I have in mind. Serial behavior of this kind is characteristic of slow, conscious thought. And Hofstadter (1985, 632) reports Simon as asserting, "Everything of interest in cognition happens above the 100 millisecond level—the time it takes you to recognise your mother." Hofstadter disagrees vehemently, asserting that everything of interest in cognition takes place below the 100-millisecond level. My position, outlined in detail in chapters 5 to 9 below, is sympathetic to Hofstadter's (and indeed owes a great deal to it). But I will show that the notion of a dispute over the correct level of interest here is misplaced. There are various explanatory projects here, all legitimate. Some require us to go below the 100-millisecond level (or whatever) while others do not. This relates, in a way I expand on later, to a problem area cited by Simon in a recent lecture (1987; see also Langley, et al. 1987, 14–16). Simon notes that programs like BACON are not good at very ill structured tasks, tasks demanding a great deal of general knowledge and expectations. Thus, though BACON neatly arrives at Kepler's third law when given the well-structured task of finding the invariants in the data, it could not come up with the flash of insight by which Fleming could both see that the mould on his petri dish was killing surrounding bacteria and recognize this as an unusual and potentially interesting event.

Listening to Simon, one gets the impression that he believes the way to solve these ill-structured problems is to *expand* the set of high-level data and heuristics that a system manipulates in the normal, slow, serial way (i.e., by creating, modifying, and comparing high-level symbol strings according to stored rules). Thus, in a recent coauthored book he dismisses the idea that the processes involved in the flash-of-insight type of discovery might be radically different in computational kind, saying that the speed and subconscious nature of such events "does not in any way imply that the process is fundamentally different from other processes of discovery—only that we must seek for other sources of evidence about its nature (i.e., subjects' introspections can no longer help)" (Langley et al. 1987, 329).

The position I develop holds rather that the folk-psychological term "scientific discovery" encompasses at least two quite different kinds of processes. One is a steady, Von Neumann-style manipulation of standard symbolic atoms in a search for patterns of regularity. And this is well modeled in Simon and Langley's work. The other is the flash-of-insight type of recognition of something unusual and interesting. And this, I shall suggest, may require modeling by a method quite different (though still computational).

In effect, theorists such as Langley, Simon, Bradshaw, and Zytkow are betting that *all* the aspects of human thought will turn out to be dependent on a single kind of computational architecture. That is an architecture in which data is manipulated by the copying, reorganizing, and pattern matching capabilities deployed on list structures by a Von Neumann (serial) processor. The basic operations made available in such a setup *define* the computational architecture it is. Thus, the pattern matching operations which such theorists are betting on are the relatively basic ones available in such cases (i.e., test for complete syntactic identity, test for syntactic identity following variable substitution, and so on). Other architectures (for example, the PDP architecture discussed in part 2 of this book) provide different basic operations. In the case of parallel distributed processing these include a much more liberal and flexible pattern-matching capacity able to find a best match in cases where the standard SPSS approach would find no match at all (see especially chapters 6 and 7 below).

Langley, Simon, et al. are explicit about their belief that the symbol-processing architecture they investigate has the resources to model and explain all the aspects of human thought. Faced with the worry that the approach taken by BACON, DALTON, GLAUBER, and STAHL won't suffice to explain all the psychological processes that make up scientific discovery, they write, "Our hypothesis is that the other processes of scientific discovery, taken one by one, have [the] same character, so that programs for discovering research problems, for designing experiments, for designing instruments and for representing problems will be describable by means of the same kinds of elementary information processes that are used in BACON" (1987, 114). They make similar comments concerning the question of mental imagery (p. 336). This insistence on a single architecture of thought may turn out to be misplaced. The alternative is to view mind as a complex system comprising many virtual architectures. If this is true, psychological explanation will likewise need to deal in a variety of types of models, availing itself in each case of different sets of basic operations (relative to the virtual architecture).

Finally, a word about the methodology of BACON and its classical-cognitivist cousins. These programs characteristically attempt to model *fragments* of what we might term recent human achievements. By this I mean they focus on tasks that we intelligent, language-using human beings perform (or at least think we perform) largely by conscious and deliberate efforts. Such tasks tend to be well structured in the sense of having definite and recognizable goals to be achieved by deploying a limited set of tools (e.g., games and puzzles with prescribed legal moves, theorem proving, medical diagnosis, cryparithmetic and so on). They also tend to be the tasks we do slowly and badly in comparison with perceptual and sensorimotor tasks, which we generally do quickly and fluently. Some AI workers are dubious about this choice of task domain and believe it essential to tackle the fluent, unconscious stuff first before going on to model more evolutionarily recent achievements. Marr (1977, 140) gives the classic statement of this: "Problem-solving research has tended to concentrate on problems that we understand well intellectually but perform poorly on.... I argue that [there are] exceptionally good grounds for *not* studying how we carry out such tasks yet. I have no doubt that when we do (e.g.) mental arithmetic we are doing *something* well, but it is not arithmetic and we seem far from understanding even one component of what that something is. Let us therefore concentrate on the simpler problems first." I have expressed similar views based on direct evolutionary arguments (Clark 1986). There still seems to me to be much truth in such strictures. But the overall picture to be developed here is rather more liberal, as we shall see.

5 *Semantically Transparent Systems*

It is time to expand on the notion of a standard symbolic atom, introduced in section 3 above. One of the most theoretically interesting points of contrast between classical systems (as understood by philosophers like Fodor and Pylyshyn) and connectionist systems (as understood by theorists like Smolensky) concerns the precise sense in which the former rely on, and the latter eschew, the use of such symbolic atoms. To bring out what is at issue here, I shall speak of the classicist as (by definition) making a methodological commitment to the construction of semantically transparent systems. Credit for the general idea of semantic transparency belongs elsewhere. The analysis I offer is heavily influenced by ideas in Smolensky 1988 and Davies, forthcoming.[7]

> A system will be said to be *semantically transparent* just in case it is possible to describe a neat mapping between a symbolic (conceptual level) semantic description of the system's behavior and some *projectible* semantic interpretation of the internally represented *objects* of its formal computational activity.

The definition is somewhat complex and is not expected to make immediate sense. In particular, the notion of a projectible semantic interpretation must remain as a kind of dummy until much later (chapter 6, section 3). It should be possible, however, to make some sense of the overall drift of the definition immediately.

The general notion of a semantically transparent system (STS) may be best appreciated from the perspective offered by Marr's now-standard account of the levels of understanding of an information-processing task. Marr (1982) distinguishes three levels at which a machine carrying out an information-processing task needs to be understood.

> *Level 1, computational theory.* This level describes the goal of the computation, the general strategies for achieving it, and the constraints on such strategies.
>
> *Level 2, representation and algorithm.* This describes an algorithm, i.e., a series of computational steps that does the job. It also includes details of the way the inputs and outputs are to be represented to enable the algorithm to perform the transformation.
>
> *Level 3, implementation.* This shows how the computation may be given flesh (or silicon) in a real machine.

In short, level 1 considers what function is being computed (at a high level of abstraction), level 2 finds a way to compute it, and level 3 shows how that way can be realized in the physical universe.

Suppose that at level 1 you describe a task by using the conceptual apparatus of public language. (This is not compulsory at level 1 but is often the case.) You might use such words as "liquid," "flow," "edge," and so on. You thus describe the function to be computed in terms proper to what Paul Smolensky calls the conceptual level, the level of public language. Very roughly, a system will count as an STS if the computational objects of its algorithmic description (level 2) is isomorphic to its task-analytic description couched in conceptual level terms (level 1). What this means is that the computational operations specified by the algorithm are applied to internal representations that are projectibly interpretable as standing for conceptual-level entities. (Again, clarification of the notion of projectibility will have to wait until chapter 6.)

Some examples may help to sharpen these levels. Consider the following specifications of functions to be computed.

(1) If (cup and saucer) then (cup)
If (cup and saucer) then (saucer)
(2) If (verb stem + ending) then (verb stem + -ed)

The functions in (1) are clear examples of a conceptual level specification. Though (2) does not draw on daily language, it is nonetheless a related case, as we shall see in subsequent chapters. In each case the items in parentheses are structural descriptions whose structure is semantically significant.

A semantically transparent system, we may now say, is one in which the *objects* (e.g., "cup and saucer") of state-transition rules in the task analysis (e.g., the rule "if (cup and saucer) then (cup)") have *structural analogues* in the actual processing story told at level 2. That is to say, in the case of (1), the level 2 story will involve computational operations defined to apply to representations sharing the complex structure of the expression "cup and saucer." In the case of (2), the level 2 story will involve computational operations defined to apply to descriptions of input verbs in a way that reveals them to have the structure "verb stem + ending." It is in this sense that classical, semantically transparent systems may be said to have a certain kind of *syntax*. For they posit mental representations that have actual structures echoing the semantic structures of our level-1 description. As Fodor and Pylyshyn (1988) forcibly point out, this is very handy if we want our system to perform systematically with respect to a certain semantic description. For it has the effect of making the semantic description a real object for the system. Hence, any inferences, etc., that are systematic in the semantic description can easily be mimicked by relying on the syntactic properties of its internal representations. If we want our system to treat "(cup and saucer)" as an instance of a general logical schema "(*a* and *b*)" and hence to perform all kinds of deductive inferences on arguments involving cups and saucers, this will be a simple matter just so long as the system's representation of cups and saucers is semantically transparent and preserves structure.

Clearly, the notion of a semantically transparent system is intended to capture the substance of Fodor and Pylyshyn's definition of a classical approach to cognitive science. Classical and connectionist approaches differ, according to Fodor and Pylyshyn, in two vital respects.

(1) "Classical theories—but not connectionist theories—posit a 'language of thought'." This means that they posit mental representations (data structures) with a certain form. Such representations are *syntactically structured*, i.e., they are systematically built by combining atomic constituents into molecular assemblies, which (in complex cases) make up whole data structures in turn. In short, they posit *symbol systems* with a combinatorial syntax and semantics.

(2) "In classical models, the principles by which mental states are transformed, or by which an input selects the corresponding output, are defined over structural properties of mental representations. Because classical mental *representations* have combinatorial structure, it is possible for classical mental *operations* to apply to them by reference to their form." This means that *if* you have a certain kind of structured representations available (as demanded by point 1), it is possible to define computational operations on those representations so that the operations are sensitive to that structure. If the structure isn't there (i.e., if there is no symbolic representation), you couldn't to it, though you might make it *look* as if you had by fixing on a suitable function in extension. (Quotes are from Fodor and Pylyshyn 1988, 12–13.)

In short, a classical system is one that posits syntactically structured, symbolic *representations* and that defines its computational *operations* to apply to such representations in virtue of their structure.

The notion of a semantically transparent system is also meant to capture the spirit of Smolensky's views on the classical/connectionist divide, as evidenced in comments like the following.

> A symbolic model is a *system* of interacting processes, all with the same conceptual-level semantics as the task behavior being explained. Adopting the terminology of Haugeland (1978), this *systematic explanation* relies on a *systematic reduction* of the behavior that involves no shift of semantic domain or *dimension.* Thus a game-playing program is composed of subprograms that generate possible moves, evaluate them and so on. In the symbolic paradigm these systematic reductions play the major role in explanation. The lowest level processes in the systematic reduction, still with the original semantics of the task domain, are then themselves reduced by *intentional instantiation*: they are implemented exactly by other processes with different semantics but the same form. Thus a move-generation subprogram with game semantics is instantiated in a system of programs with list-manipulating semantics. (Smolensky 1988, 11)

Before leaving the subject of STSs, it is worth pausing to be quite explicit about one factor that is *not* intended as part of the definition of an STS. Under the terms of the definition an STS theorist is *not* committed to any view of how the system explicitly represents the *rules* adduced in task analysis (level 1). Thus, in my example (1), there is no suggestion that the rule "If (cup and saucer) then (cup)" must *itself* be explicitly represented by the machine. A system could be an STS and be hard-wired so as to take input "cup and saucer" and transform it into output "cup." According to STS theory, all that must be explicit is the structured description of the objects to which the rule is defined to apply. The derivation rules may be tacit, so long as the data structures they apply to are explicit. On this Fodor and Pylyshyn rightly insist: "Classical machines can be *rule implicit* with respect to their programs.... What *does* need to be explicit in a classical machine is not its program but the symbols that it writes on its tapes (or stores in its registers). These, however, correspond not to the machine's rules of state transition but to its data structures" (1988, 61). As an example they point out that the grammar posited by a linguistic theory need not be explicitly represented in a classical machine. But the *structural descriptions of sentences* over which the grammar is defined (e.g., in terms of verb stems, subordinate clauses, etc.) must be. Attempts to characterize the classical/connectionist divide by reference to explicit or non explicit rules are thus shown to be in error (see also Davies, forthcoming).

6 *Functionalism*

While I am setting the stage, let me bring on a major and more straightforwardly philosophical protagonist, the functionalist. The functionalist is in many ways the natural bedfellow of the proponent of the physical-symbol-system hypothesis. For either version of the physical-symbol-system hypothesis claims that what is essential to intelligence and thought is a certain capacity to manipulate symbols. This puts the essence of thought at a level independent of the physical stuff out of which the thinking system is constructed. Get the symbol manipulating capacities right and the stuff does not matter. As the well-known blues number has it, "It ain't the meat, it's the motion." The philosophical doctrine of functionalism echoes this sentiment, asserting (in a variety of forms) that mental states are to be identified not with, say, physicochemical states of a being but with more abstract organizational, structural, or informational properties. In Putnam's rousing words "We could be made of swiss cheese and it wouldn't matter" (1975, 291). Aristotle, some would have it, may have been the first philosophical functionalist. Though there seems to be a backlash now underway (see, e.g., Churchland 1981 and Churchland 1986), the recent popularity of the doctrine can be traced to the efforts of Hilary Putnam (1960, 1967), Jerry Fodor (1968), David Armstrong (1970) and, in a slightly different vein, Daniel Dennett (1981) and William Lycan (1981). I shall not attempt to do justice to the nuances of these positions here. Instead, I shall simply characterize the most basic and still influential form of the doctrine, leaving the search for refinements to the next chapter. First, though, a comment on the question to which functionalism is the putative answer.

In dealing with the issues raised in this book, it seemed to me to be essential to distinguish the various explanatory projects for which ideas about the mind are put forward. This should become especially clear in the closing chapters. For now I note that one classical *philosophical* project has been to formulate and assess *schemas* for a substantial theory of the essence of the mental. The notion of essence here may be unpacked as the search for the necessary and sufficient conditions for being in some mental state. In this restricted sense a theory of mind should tell us what it is about a being that makes it true to assert of that being that it is in a given mental state (e.g., believing it is about to rain, feeling sad, feeling anxious, suffering a stabbing pain in the left toe, and so forth).

For the moment let me simply assert that Newell and Simon's intended project (in common with a lot of workers in AI) is psychological explanation. Pending a fuller account of psychological explanation, it is not obvious that the project of psychological explanation is identical with the project of seeking the essence of the mental in the sense just sketched. Newell and Simon's talk of the physical-symbol-system hypothesis as an account of the necessary and sufficient conditions of intelligent action effectively identifies the tasks. It follows that having a full psychological explanation in their sense would put you in a position to re-create or instantiate the analyzed mental state in a machine (barring practical difficulties). I shall later argue for a firm distinction between these projects of psychological explanation and psychological instantiation.

Functionalism, then, is a sketch or schema of the kind of theory that, when filled in, will tell us in a very deep sense what it *is* to be in some mental state. The most basic form of such a theory is known as Turing-machine functionalism. Not surprisingly, the doctrine takes its cue from Turing's conception of the formal properties sufficient to guarantee that a task is computable by a mechanism, regardless of the physical stuff out of which the mechanism was made (see section 2 above).

In Putnam's hands (1960, 1967) functionalism came to suggest a theory of mind (in the sense of a schema for a substantial theory of the essence of the mental) that was apparently capable of avoiding many of the difficulties that beset other such proposals. Very sketchily, the situation was something like this. Dualism (the idea that mind is a ghostly kind of nonmaterial *substance*) had been discredited as nonexplanatory mysticism and was briefly displaced by behaviorism. Behaviorism (Ryle 1949) held that mental states were identical with sets of actual and counterfactual overt behaviors and that inner states of the subject, though no doubt causally implicated in such behaviors, were not theoretically important to understanding what it *is* to be in certain mental states.

This dismissal of the importance of internal states (for a philosophical theory of mind) was resisted by the first wave of identity theories which claimed that mental states were identical with brain processes (Smart 1959). But the identity theory, if one took the claims of its proponents rather literally (more literally, I am inclined to think, than they ever intended) lay open to a variety of criticisms. Especially relevant here is Putnam's (1960, 1967) criticism that identity theory makes far too tight the tie between being in a certain mental state (e.g., feeling pain) and being in a certain physicochemical or neural state. For on an extreme, type-type-identity reading, the identity of some mental state with, say, some neural state would seem to imply that a being incapable of being in that neural state could not, in principle, be in the mental state in question. But for rather obvious reasons this was deemed unacceptable. A creature lacking neurons would be unable to occupy any neural state. But couldn't there be exotic beings made of other stuff who were nonetheless capable of sharing our beliefs, desires, and feelings? If we allow this seemingly sensible possibility then we, as philosophers, need some account of what physically variously constituted feelers and believers have in common that makes them feelers and believers. Behaviorism would have done the trick, but its denial of the importance of inner states had been perceived as a fault. Identity theory, it seemed, had gone too far in the other direction.

Between the Scylla and the Charybdis sailed the good ship functionalism. What is essential to being in a certain mental state, according to the functionalist schema, is being in a certain abstract functional state. And that functional state is defined over two components: (1) the role of some internal states in mediating system input and system output (the behavior element) and (2) the role of the states or processes in activating or otherwise affecting other internal states of the system (the inner element). If we *also* presume that cognition is a computational phenomenon, then we can link this characterization (as Putnam [1960] did) to the notion of a Turing machine, which is defined by its input and output and its internal state-transition profile. What Turing machine you are instantiating, not what substance you are made of, characterizes your mental states. As I said, it ain't the meat, it's the motions.

Now the bad news. Functionalism has had its problems. Of special interest to us will be the problem of excessive liberalism (see Block 1980). The charge is that Turing-machine functionalism allows too many kinds of things to be possible believers and thinkers. For example, it might in principle be possible to get the population of China to pass messages (letters, values, whatever) between themselves so as briefly to realize the functional specification of some mental state (Block 1980, 276–278). (Recall, it is only a matter of correctly organizing inputs, outputs, and internal state transitions, and these, however they are specified, won't be tied to any particular kind of organism.) As Block (1980, 277) puts it, "In describing the Chinese system as a Turing machine I have drawn the line [i.e., specified what counts as inputs and outputs] in such a way that it satisfies a certain type of functional description—one that you *also* satisfy, and one that, according to functionalism, justifies attributions of mentality." But, says Block, there is at least prima facie reason to doubt that such a system could have any mental states at all. Could that overall system really constitute, say, an agent in pain? Surely not. Surely there is nothing which it is like, either nice or nasty, to be such a system. It has no phenomenal or subjective experience. Or as philosophers put it, the system has no qualia (raw feels, real subjectivity). Hence, Block dubs this argument the absent-qualia argument. I suspect that this argument loses much of its force once the functionalist hypothesis is firmly disassociated from both STS-and SPSS-type approaches (see chapters 2 and 10). But for the moment, I leave the discussion of functionalism on that quizzical note.

The players are on stage. We have met the physical-symbol-system hypothesis and its methodological cousin, semantically transparent cognitive modeling. We have met a computationally inspired philosophical model of mind (functionalism) and hinted at its difficulties. The disembodied presence of connectionism awaits future flesh. But we should first pause to see what philosophers have made of the story so far.

Chapter 2
Situation and Substance

1 *Stuff and Common Sense*

Classical cognitivism was not universally celebrated by philosophers. There was a slightly inchoate sense of mind being finessed by sleight of hand. With the work of Dreyfus (1972, 1981) and later Searle (1980), the inchoate became flesh. The flesh, however, was quite different in each case. For Dreyfus the problem with classical cognitivism was the adequacy of current (and perhaps any) representational formalisms to re-create and express human commonsense knowledge. For Searle the worry was a perceived gap between the formal and the intentional, or between syntax and meaning. The biological stuff of which we are made, Searle felt, was at least as essential to our abilities to know and understand as any formal or computational properties it may happen to possess. In this chapter I shall examine these two kinds of worry, exposing a few of the threads I later weave into a defence of cognitive science.

2 *The Dreyfus Case*

Hubert Dreyfus has been one of the most persistent yet sensitive critics of the cognitivist tradition. At the core of his disquiet lies the thought that there is a richness or "thickness" to human understanding that cannot be captured in any set of declarative propositions or rules. Instead, the richness depends on a mixture of culture, context, upbringing, and bodily self-awareness. Accordingly, Dreyfus turns a sceptical gaze on the early microworlds work associated with Winograd (1972) and the frame- and script-based approaches associated with Minsky (1974) and Schank and Abelson (1977), among others.

Winograd's (1972) SHRDLU was a program that simulated a robot acting in a small microworld composed entirely of geometric solids (blocks, pyramids, etc.). By restricting the domain of SHRDLU's alleged competence and providing SHRDLU with a model of the items in the domain, Winograd was able to produce a program that could engage in a quite sensible dialogue with a human interlocutor. SHRDLU could, for example, resolve problems concerning the correct referents of words like "it" and "the pyramid" (there were many pyramids) by deploying its knowledge of the domain. The output of the program was relatively impressive, but its theoretical significance as a step on the road to modeling human understanding was open to doubt. Could a suitable extension of the microworld strategy really capture the depth and richness of human understanding? Dreyfus thought not, since he saw no end to the number of such micro competences required to model even a child's understanding of a real-world activity like bargaining.

Consider the following example (due to Papert and Minsky and cited in Dreyfus 1981, 166).

> *Janet:* "That isn't a very good ball you have. Give it to me and I'll give you my lollipop."

For the set of micro theories needed to understand Janet's words, Minsky and Papert suggest a lengthy list of concepts that includes: time, space, things, people, words, thoughts, talking, social relations, playing, owning, eating, liking, living, intention, emotions, states, properties, stories, and places. Each of these requires filling out. For example, anger is a state caused by an insult (and other causes) and results in noncooperation (and other effects). And this is just one emotion. This is a daunting list and one that, as Dreyfus points out, shows no obvious signs of completion or even completability. Dreyfus thus challenges the micro theorist's faith that some finite and statable set of micro competences can turn the trick and result in a computer that really knows about bargaining or anything else. Such faith, he suggests, is groundless, as AI has failed "to produce even the hint of a system with the flexibility of a six-month old child" (Dreyfus 1981, 173). He thinks that "the special purpose techniques which work in context-free, gamelike micro-worlds may in no way resemble general purpose human and animal intelligence."

Similar criticisms have been applied to Winston's (1975) work in computer vision and Minsky's (1974) frame-based approach to representing everyday knowledge. A frame is a data structure dealing with a stereotypical course of events in a given situation. It consists of a set of modes and relations with slots for specific details. Thus a birthday-party frame would consist of a rundown of a typical birthday-party-going sequence and might refer to cakes, candles, and presents along the way. Armed with such a frame, a system would have the common sense required to assume that the cake had candles, unless told otherwise.

But as Dreyfus is not slow to point out, the difficulties here are still enormous. First, the frames seem unlikely ever to cover all the contingencies that common sense copes with so well. (Are black candles on a birthday cake significant?) Second, there is the problem of accessing the right frame at the right time. Humans can easily make the transition from, say, a birthday-party frame to an accident frame or to a marital-scene frame. Is this done by yet more, explicit rules? If so, how does the system know when to apply these. How does it tell what rule is relevant and when? Do we have a regress here? Dreyfus thinks so. To take one final example, consider the attempt to imbue an AI system with knowledge of what it is for something to be a chair. To do so, Minsky suggests, we must choose a set of chair-description frames. But what does this involve? Presumably not a search for common physical features, since chairs, as Dreyfus says, come in all shapes and sizes (swivel chairs, dentist's chairs, wheelchairs, beanbag chairs, etc.). Minsky contemplates functional pointers, e.g., "something one can sit on." But that is too broad and includes mountains and toilet seats. And what of toy chairs, and glass chairs that are works of act? Dreyfus' suspicion is that the task of formally capturing all these inter-linked criteria and possibilities in a set of context-free facts and rules is endless. Yet, as Dreyfus can hardly deny, human beings accomplish the task somehow. If Dreyfus doubts the cognitivist explanation, he surely owes us at least a hint of a positive account.

3 *It Ain't What You Know; It's the Way You Know It*

And we get such an account. In fact, we get hints of what I believe to be two distinct kinds of positive account. Critics of Dreyfus (e.g., Torrance [1984, 22–23]) have tended to run these together, thereby giving the impression that reasonable doubts concerning one line of Dreyfus's thought constitute a global undermining of his position. In this section I shall try to be a little more sympathetic. This is not to say, however, that such critics are wrong or lack textual evidence for their position. In early articles Dreyfus does indeed seem to run together the two accounts I distinguish. And even in the most recent book (Dreyfus and Dreyfus 1986), he treats the two as being intimately bound up. My claim is that they both can and *should* be kept separate.

The first line of thought concerns what I shall call "the body problem." The body problem is neatly summed up in the following observation: "The computer comes into our world even more alien than would a Martian. It does not have a body, needs or emotions, and it is not formed by a shared language and other social practices" (Dreyfus and Dreyfus 1986, 79). We can see this worry at work in the following passage, in which Dreyfus tries for a positive account of what makes something a chair: "What makes an object a chair is its function [and] its place in a total practical context. This pre-supposes certain facts about human beings (fatigue, the way the body bends) and a network of other culturally determined equipment (tables, floors, lamps) and skills (eating, writing, going to conferences, giving lectures etc.).... Moreover, understanding chairs also includes social skills such as being able to sit appropriately... at dinner, interviews, desk jobs" Dreyfus (1981, 184). Such comments stress the role of the human body and our slow social and cultural upbringing, our "slowly acquired situated understanding." Now certainly our bodies and social and cultural contexts constitute a rich source of knowledge about the human world. Thus, suppose we want a computer to parse (or assign grammatical categories and values) and understand the instructions on a bottle of shampoo. On one brand in my household these include: "As with all shampoos, avoid getting this shampoo into the eyes. If it does, rinse well with warm water."[1] Imagine trying to program a bodyless machine successfully to parse such a message! Out of context it is not *at all* obvious *what* the second sentence would have us rinse. We understand the message in part because we have eyes and we know they are easily irritated and that rinsing the eyes with water helps. This is just one example of a host of knowledge that we are privy to simply in virtue of having the bodies and the bodily reactions we do. The attempt to force-feed such knowledge (along with whatever we pick up socially and so on) into a nonsocial, bodyless machine is almost certain to fail by falling far short of the richness of adult human understanding.

The difficulty is knowing exactly what theoretical force attaches to such observations. For they suggest a reading of Dreyfus that seems to saddle him with a rather implausible critical-mass argument, which might go something like this.

> Until you know a certain *amount* about something, you don't *really* know anything about it at all.
>
> This critical mass is attained only through our bodily, social and cultural interactions with the world.
>
> So a computer, lacking our socially situated and embodied advantages, can never know anything.

Alas, Dreyfus sometimes gives us good cause to ascribe some such argument to him. Speaking of a program (see section 4 below) that seeks to model knowledge about restaurants he writes, "When the waitress came to the table, did she wear clothes? Did she walk forwards or backwards? Did the customer eat his food with his mouth or his ear? If the program answers, 'I don't know' we feel that all of its right answers were tricks or lucky guesses and that it has not understood anything of our everyday restaurant behaviour" (Dreyfus 1981, 189). This "If you don't know *that*, you don't know nothin'" approach has struck many critics as totally implausible.[2] We often (correctly, I believe) ascribed *some* understanding of things to human children in the face of the most bizarre misconceptions and gaps in their knowledge. And couldn't a Martian be said to have learned something of eating by learning what we eat and why, even before knowing enough about human anatomy to decide whether we eat with our mouths or our ears? Critical-mass arguments thus strike me as very unconvincing. Let us therefore pass quickly on to another line of thought.

The second line of Dreyfus's thought takes its cue not from the body problem but from more general observations about ways of encoding knowledge (if "encoding" is the right word). It is here that his ideas are most suggestive. He notes that human beings seem naturally inclined to spot the significant aspects of a situation, and he relates this capacity not to a stored set of rules and propositions but to their vast experience of previous concrete situations and some kind of holistic associative memory. He asks, "Is the know-how that enables humans constantly to sense what specific situations they are in the sort of know-how that can be represented as a kind of knowledge in *any* knowledge representation language no matter how ingenious and complex?" (Dreyfus 1981, 198.) Viewed from this angle, Dreyfus's worry is *not* that machines don't know enough (because of their lack of bodies and so on) but rather that the *way* in which current AI programs represent knowledge is somehow fundamentally inadequate to the real task. Such programs assume what Dreyfus doubts, namely, that "all that is relevant to intelligent behaviour can be formalised in a structured description" (p. 200). This is, for Dreyfus, the most basic tenet of what he calls the "information processing approach."

At this point I must pause to raise a few questions of my own. What exactly counts as a structured description? What counts as a knowledge-representation language? It may seem as if Dreyfus here intends to rule out *all* forms of computational accounts of cognition and lay total (and mysterious) stress on human bodies and culture. This, however, is not the case. In a recent book Dreyfus stresses the need for flexible systems capable of what he calls "holistic similarity recognition" if any progress is to be made in modeling human expertise (Dreyfus and Dreyfus 1986, 28). He cites, with some approval, recent work on connectionist or PDP approaches to mind (p. 91). Perhaps most revealingly of all, he does so in a section entitled "AI without Information Processing." As we shall see, I disagree with the claim that such approaches do away with structured descriptions or information processing. But that, for now, is another matter. What we need to notice here is just that since Dreyfus's doubts exclude such approaches but include the approaches taken by Newell, Simon, Winograd, and others, they may best be seen as doubts about what I earlier dubbed the SPSS hypothesis. They are doubts about whether a *certain* computational approach can in principle yield systems with the kind of flexibility and common sense we tend to associate with the warranted ascription of understanding. The underlying thought, in effect, is that where real intelligence is concerned, it ain't *what* you know, it's the *way* you know it. Of course, there could be a link with the points about bodies and so forth even here. There may be some things we know about largely by our own awareness of our bodily and muscular responses (Dreyfus cites swimming as an example). Perhaps a machine lacking our kind of body but equipped with some kind of mechanism for holistic similarity processing could even so know less about such things than we can. Nonetheless, on the reading of Dreyfus I am proposing (I have no idea whether he would endorse it), such a machine would be flexible and commonsensical within its own domains, and as such it would be at least a candidate for a genuine knower, albeit not quite a human one. This is in contradistinction to any system running a standard cognitivist program. I shall expand on this point in subsequent chapters.

Of Dreyfus's two points (the one about the social and embodied roots of human knowing, the other about the need for flexible, commonsense knowledge) it is only the second which I think we can expect to bear any deep theoretical weight. But at least that point is suggestive. So let us keep it in our back pockets as we turn to a rather different criticism of AI and cognitive science.

4 *Manipulating the Formal Shadows of Mind*

In a series of recent publications (1980, 1983, 1984) John Searle has established himself as a leading opponent of the information-theoretic approach to mind. That approach, he thinks, is just tilting at the "formal shadows" of mind. But in contrast, real mentality, he says, depends on far wetter things, namely, on the physicochemical properties of human brains. Searle's criticism is targeted on what he calls "the hypothesis of strong AI." This is defined as the claim that "the appropriately programmed computer literally has cognitive states and that the programs thereby explain human cognition" (Searle 1980, 283). The attack begins with a now infamous thought experiment, the puzzling case of the Chinese room. This thought experiment aims to provide a general critique of the computational approach to mind. Its starting point is a specific program that might seem to simulate the intentional activity of understanding a story (Schank and Abelson 1977). Very briefly, the program provides the computer with some background data concerning the topic of a story to be presented. The computer can then be given a story on this topic and afterward it will answer questions about the story that are not explicitly resolved in the story itself. Thus, to use Searle's example (which I touched on earlier), we might program in background data on human behavior in restaurants. We may then tell the story of a man who enters a restaurant, orders a hamburger, and upon leaving, presents the waitress with a big tip. If the computer is then asked, "And did the man *eat* the hamburger?" it can answer "yes," because it apparently knows about restaurants. Searle believes, I think rightly, that the computer running this program does not really know about restaurants at all, at least if by "know" we mean anything like "understand." The Chinese-room example is constructed in part to demonstrate this. But Searle believes his arguments against *that* sort of computational model of understanding are also arguments against *any* computational model of understanding.

We are asked to imagine a human agent, an English monolinguist, placed in a large room and given a batch of papers with various symbols on it. These symbols, which to him are just meaningless squiggles identifiable only by shape, are in fact the ideograms of the Chinese script. A second batch of papers arrives, again full of ideograms. Along with it there arrives a set of instructions in English for correlating the two batches. Finally, a third batch of papers arrives bearing still further arrangements of the same uninterpreted formal symbols and again accompanied by some instructions in English concerning the correlation of this batch with its predecessors. The human agent performs the required matchings and issues the result, which I shall call "the response." This painstaking activity, Searle argues, corresponds to the activity of a computer running Schank's program. For we may think of batch 3 as the questions, batch 2 as the story, and batch 1 as the script or background data. The response, Searle says, may be so convincing as to be indistinguishable from that of a true Chinese speaker. And yet, and this is the essential point, the human agent performing the correlations understands no Chinese, just as, it would now appear, a computer running Schank's program understands no stories. In each case what is going on is the mere processing of information. If the intuitions prompted by the Chinese-room example are correct, understanding must involve something extra. From this Searle concludes that no computer can ever understand merely by "performing computational operations on formally specified elements." Nor, consequently, can the programs that determine such computational operations tell us anything about the special nature of mind (Searle 1980, 286).

Ramming the point home, Searle asks us to compare the understanding we (as ordinary English speakers) have of a story in English against the "understanding" the person manipulating the formal symbols in the Chinese room has of Chinese. There is, Searle argues, no contest. "In the Chinese case I have everything that Artificial Intelligence can put into me by way of a program and I understand nothing; in the English case I understand everything and there is so far no reason at all to suppose that my understanding has anything to do with computer programs—i.e., with computational operations on purely formally specified elements" (Searle 1980, 286). In short, no formal account can be *sufficient* for understanding, since "a human will be able to follow the formal principles without understanding anything" (p. 287). And there is no obvious reason to think that satisfying some *formal* condition is *necessary* either, though as Searle admits, this could (just conceivably) yet prove to be the case. The formal descriptions, Searle thinks (p. 299), seem to be capturing just the shadows of mind, shadows thrown not by abstract computational sequences but by the actual operation of the physical stuff of the brain.

I shall argue that Searle is simply wrong thus *completely* to shift the emphasis away from formal principles on the basis of a demonstration that the operation of a certain *kind* of formal program is insufficient for intentionality. The position to be developed below and in chapters 3 and 5 to 11 views as a necessary though perhaps insufficient condition of real understanding the instantiation of a certain kind of formal description that is far more microstructural than the descriptions of the SPSS hypothesis. Undermining Searle's strongest claims, however, is no simple matter, and we must proceed cautiously. The best strategy is to look a little more closely at the *positive* claims about the importance of the nonformal, biological stuff.

5 *Showing What We're Made Of*

Searle considers several possible replies to his paper, only one of which need interest us here.[3] It is what he calls the brain-simulator reply, and it goes like this. Suppose a program modeled the formal structure of actual Chinese brains engaged in understanding Chinese. Surely *then* it would constitute a genuine Chinese understanding. At this point, to his credit, Searle grasps the nettle. "No," he says, "we could imagine an elaborate set of water pipes and valves, and a human switcher, realising *that* formal description too. But wherein would the understanding of Chinese reside? Surely the answer is 'nowhere'" (adapted from Searle 1980, 295). (This argument should recall the worries about excessive liberalism and absent qualia, which, at the last showing (chapter 1), had functionalism in a vicelike grip.) Regarding the brain simulator, then, Searle is in no doubt: "As long as it simulates only the formal structure of the sequence of neuron firings at the synapses, it won't have simulated what matters about the brain, namely its causal properties, its ability to produce intentional states" (Searle 1980, 295). Or again: "What matters about brain operation is not the formal shadow cast by the sequence of synapses but rather the actual properties of the sequences" (Searle 1980, 300).

The allusions to causal powers have struck many critics as unforgivably obscure. It is hard to see why. Searle's claim has two components: (1) The formal properties of the brain do not constitute intentionality. (2) the reason they do not constitute it is that only certain kinds of stuff can support thought. Well, (1) may be right (see chapter 3), though *not* for the reasons cited in (2). But even so, (2) is surely not *that* obscure a claim. Searle cites the less puzzling case of photosynthesis. By focusing on this, we may begin to unscramble the chaos.

Photosynthesis, Searle suggests, is a phenomenon dependent on the actual causal properties of certain substances. Chlorophyl is an earthly example. But perhaps other substances found elsewhere in the universe can photosynthesize too. Similarly, Martians might have intentionality, even though (poor souls) their brains are made of different stuff from our own. Suppose we now take a formal chemical theory of how photosynthesis occurs. A computer could then work through the formal description. But would actual photosynthesis thereby take place? No, it's the wrong *stuff*, you see. The formal description is doubtless a handy thing to have. But if it's *energy* (or thought) you need, you had better go for the real stuff. In its way, this is fair enough. A gross formal theory of photosynthesis might consist of a single production, "If subjected to sunlight, then produce energy." A fine-grained formal theory might take us through a series of microchemical descriptions in which various substances combine and cause various effects. Gross or fine-grained, neither formalism seems to herald the arrival of the silicon tulip. Market gardening has nothing to fear from simulated gardening as yet.

Now, there are properties of plants that are irrelevant to their photosynthetic capacities, e.g., the color of blooms, the shape of leaves (within limits) the height off the ground, and so on. The questions to ask are: *What* do the chemical properties buy for the plant, and what are the properties of the chemicals by which they buy it? The human brain is made out of a certain physical, chemical stuff. And perhaps in conjunction with other factors, that stuff buys us thought, just as the plant's stuff buys it energy. So, what are the properties of the physical chemical stuff of the brain that buy us thought? Here is one answer (not Searle's or that of supporters of Searle's emphasis on stuff, e.g., Maloney [1987]): the vast structural variability in response to incoming exogenous and endogenous stimuli that the stuff in that arrangement provides.[4]

Suppose this were so. Might it not also be true that satisfying *some* kinds of formal description guaranteed the requisite structural variability and that satisfying *other* kinds of formal description did not? Such a state of affairs seems not only possible but pretty well inevitable. But if so, Searle's argument against the formal approach is, to say the least, inconclusive. For the only evidence against the claim that the formal properties of the brain buy it structural variability, which in turn buys it the capacity to sustain thought, is the Chinese-room thought experiment. But in that example the formal description was at a very gross level, in line with the SPSS hypothesis of chapter 1, which in this case amounts to rules for correlating inputs, corresponding to sentences of Chinese, with similar outputs. It could well be that a system capable of satisfying *that* level of formal description need not possess the vast structural variability by which (on my hypothesis) the brain supports thought. This could be neatly tied in with Dreyfus's observations that implementations of conventional cognitivist programs are inflexible and lack common sense. Such programs do not depend on a suitably variable and flexible substructure and hence fail to instantiate any understanding whatsoever. (If this talk about suitably variable and flexible substructures seems mysterious, it should become less so once we look at new kinds of computational models of mind: connectionist or PDP models.)

But it might yet prove to be the case that formal descriptions at a *lower*, more microstructural level will have only instantiations that *must* constitute a system with the requisite structural variability. And as long as this possibility remains, the case for the importance of stuff is far from watertight. Moreover, as we shall see in later chapters, cognitive science is just beginning to develop formal, microstructural theories that fit this general bill (see chapters 5 to 10). The price of this maneuver is, of course, grasping Searle's nettle at the other end. If a set of pipes really did constitute a system with the requisite structural variability, then (subject, perhaps, to a few further stipulations—see chapter 3) we should welcome it as a fellow thinker. I am ecumenical enough to do this. The more so since, I am reasonably convinced, it is at least physically impossible to secure the relevant variability out of such parts in the actual universe. If there are possible worlds subject to different physical laws than our own and if in those worlds collections of pipes, beer cans or whatever exhibit the relevant fine-grained formal properties (if, for example, they are organized into a value passing network with properties of relaxation, graceful degradation, generalization, and so on [see chapter 5]), we should bear them no ill will Some beer cans, it seems, satisfy formal descriptions that our beer cans cannot reach.

6 *Microfunctionalism*

The defence of a formal approach to mind mooted above can easily be extended to a defence of a form of functionalism against the attacks mounted by Block (see chapter 1, section 5). An unsurprising result, since Searle's attack on strong AI is intended to cast doubt on *any* purely formal account of mind, and that attack, as we saw, bears a striking resemblance to the charges of excessive liberalism and absent qualia raised by Block. Functionalism, recall, identified the real essence of a mental state with an input internal state transition, and output profile. Any system with the right profile, regardless of its size, nature and components, would occupy the mental state in question. But unpromising systems (like the population of China) could, it seemed, be so organized. Such excessive liberalism seemed to undermine functionalism: surely the system comprising the population of China would not *itself* be a proper subject of experience. The qualia (subjective experience or feels) seem to be nowhere present.

It is now open to us to respond to this charge in the same way we just responded to Searle. It all depends, we may say, on where you locate the grain of the input, internal state transitions, and output. If you locate it at the gross level of a semantically transparent system, then we may indeed doubt that satisfying *that* formal description is a step on the road to being a proper subject of experience. At that level we may expect absent qualia, excessive liberalism, and all the rest, although this needn't preclude formal accounts at that level being good *psychological* explanations in a sense to be developed later (chapters 7 and 10). But suppose our profile is much finer-grained and is far removed from descriptions of events in everyday language, perhaps with internal-state transitions specified in a mathematical formalism rather than in a directly semantically interpretable formalism. Then it is by no means so obvious (if it ever was—see Churchland and Churchland 1981) either that a system made up of the population of China *could* instantiate such a description or that if it did, it would not be a proper subject of the mental ascriptions at issue (other circumstances permitting—see chapter 3). My suggestion is that we might reasonably bet on a kind of microfunctionalism, relative to which our intuitions about excessive liberalism and absent qualia would show up as more clearly unreliable.

Such a position owes something to Lycan's (1981) defence of functionalism against Block. In that defense he accuses Block of relying on a kind of gestalt blindness (Lycan's term) in which the functional components are made so large (e.g., whole Chinese speakers) or unlikely (e.g., Searle's beer cans) that we rebel at the thought of ascribing intentionality to the giant systems they comprise. Supersmall beings might, of course, have the same trouble with neurons. Lycan, however, then opts for what he calls a homuncular functionalism, in which the functional subsystems are identified by whatever they may be said to do for the agent.

Microfunctionalism, by contrast, would describe at least the *internal* functional profile of the system (the internal state transitions) in terms far removed from such contentful, purposive characterizations. It would delineate formal (probably mathematical) relations between processing units in a way that when those mathematical relations obtain, the system will be capable of vast, flexible structural variability and will have the attendant emergent properties. By keeping the formal characterization (and thereby any good semantic interpretation of the formal characterization) at this fine-grained level we may hope to guarantee that any instantiation of such a description provides at least potentially the right kind of substructure to support the kind of flexible, rich behavior patterns required for true understanding. These ideas about the right kind of fine-grained substructures will be fleshed out in later chapters.

Whether such an account is properly termed a species of functionalism, as I've suggested, is open to some debate. I have opted for a broad notion of functionalism that relates the real essence of thought and intentionality to patterns of nonphysically specified internal state transitions suitable for mediating an input-output profile in a certain general kind of way. This in effect identifies functionalism with the claim that structure, not the stuff, counts and hence identifies it with *any* formal approach to mind. On that picture, microfunctionalism is, as its name suggests, just a form of functionalism, one that specifies internal state transitions at a very fine-grained level.

Some philosophers, however, might prefer to restrict the "functionalism" label to just those accounts in which (1) we begin by formulating, for *each* individual mental state, a profile of input, internal state transitions, and output in which internal state transitions are described at the level of beliefs, desires, and other mental states of folk psychology (see the next chapter) (2) we then replace the folk-psychological specifications by some formal, nonsemantic specification that preserves the boundaries of the folk-psychological specifications.[5] Now there is absolutely no guarantee that such boundaries will be preserved in a microfunctionalist account (see the next chapter). Moreover, though it may, microfunctionalism need not aspire to give a functional specification of *each* type of mental state. (How many are there anyway?) Instead, it might give an account of the kind of substructure needed to support general, flexible behavior of a kind that makes appropriate the ascription to the agent of a whole *host* of folk-psychological states. For these reasons, it may be wise to treat "microfunctionalism" as a term of art and the defence of functionalism as a defence of the possible value of a fine-grained formal approach to mind. I use the terminology I do because I believe the essential motivation of functionalism lies in the claim that what counts is the structure, not the stuff (this is consistent with its roots—see Putnam 1960, 1967, 1975b). But who wants to fight over a word? Philosophical disquiet over classical cognitivism, I conclude, has largely been well motivated but at times overambitious. Dreyfus and Searle, for example, both raise genuine worries about the kind of theories that seek to explain mind by detailing computational manipulations of standard symbolic atoms. But it is by no means obvious that criticisms that make sense relative to those kinds of computational models are legitimately generalized to *all* computational models. The claim that structure, not stuff, is what counts has life beyond its classical cognitivist incarnation, as we shall see in part 2.

Chapter 3
Folk Psychology, Thought, and Context

1 *A Can of Worms*

Imagine a can of worms liberally spiced with red herrings. Such, I believe, is the continuing debate over the role of folk-psychology in cognitive science. What faces us is a convulsing mass of intertwined but ill-defined issues like:

- Is commonsense talk of our mental lives (folk-psychology) a proto-scientific theory of the inner wellsprings of human action?
- Should we expect a neat, boundary-preserving reduction of folk-psychological talk to the categories of a scientific psychology?
- Failing such reduction, can cognitive science properly claim to be studying the *mind*?
- Conversely, could progress in cognitive science force the abandonment or revision of ordinary folk-psychological talk of beliefs and desires? Is it likely to?

There is a sprawling literature here: Churchland 1981, Stich 1983, Fodor 1980a, Searle 1983, McGinn 1982, Millikan 1986, Pettit and McDowell 1986, and Clark 1987a. And that barely scratches the surface. My strategy will be to divide and selectively ignore.

It may be helpful briefly to gesture at the relation of these issues to the overall themes of my discussion. One major goal of this book is to develop a framework in which some formal approaches to mind can be seen as plausible, despite the kinds of objections raised in the last chapter, and philosophically respectable. This demand of philosophical respectability requires me to be quite canny about the precise *way* in which formal or computational considerations are meant to illuminate the mind. One stumbling block here is the thought that the very *idea* of mind is intimately tied up with our ordinary talk of mental states like belief, desire, hope, and fear. *These* mental states, according to some of the arguments treated below, will necessarily (on some accounts) or probably (on others) elude analysis in terms of the internal states that cognitive science eventually endorses. So whatever else cognitive science does, it won't succeed (so the argument goes) in illuminating the nature of *mind*. If you accept this (which you need not do), you might either conclude: so much the worse for the commonsense idea of mind (Churchland 1981; Stich 1983), or so much the worse for the claims of cognitive science to be investigating the nature of mind (see various articles in Pettit and McDowell 1986). It thus falls to any would-be apologist for cognitive science to delve at least some way into the vermian heap. So here goes.

2 A Beginner's Guide to Folk Psychology

The good news about folk psychology is that a beginner's guide won't be necessary after all. For it is a fair bet that nobody reading this book is a beginner. The term "folk psychology" refers simply to our mundane, daily understanding of ourselves and others as believing, hoping, fearing, desiring, and so forth. Some such understanding of mental states is the common property of most adult human speakers in contemporary societies. At its core it uses belief and desire ascriptions to shed light on behavior or (to avoid begging questions) bodily movements.

A colleague suddenly gets up and rushes over to the bar. You explain her movements by saying, "She is desperate for a Guinness and believes she can get one at the bar." Very likely you won't express yourself in such a stilted and artificial way in such a simple case. But the explanatory mode is familiar enough, and it is one we explicitly use in more complex cases, e.g., solving detective mysteries that involve the search for motives for the evil deed.

At a minimum folk psychology is thus the use of belief and desire talk to explain action or, better, movement (movement becomes action when it is subsumable under the intentional umbrella of a folk-psychological understanding). Folk psychology is *not* the gossip's understanding of, e.g., Freudian theories of psychoanalysis. In that respect the term "folk pyschology" is somewhat misleading. In the recent literature (Churchland 1979, 1981; Stich 1983) folk psychology is treated (and criticized) as a primitive, protoscientific *theory* of the internal causal antecedents of human behavior. At first blush this may seem a somewhat startling thought. What could be meant by the claim that our ordinary ideas about the mental involve some kind of theory? And even if they do, why should it be a theory of *internal* causes of behavior? Let us address the first of these questions and leave the other to ferment until later. Return to the case of the lover of Guinness. Our belief-desire description of her movements toward the bar is *explanatory*, it is argued, only if we tacitly accept a general psychological law. In this case:

$$(x)(p)(q)\{[(x \text{ desires that } p) \ \& \ (x \text{ believes that } (q \rightarrow p))] \rightarrow (x \text{ will try, all else being equal, to bring it about that } q)\}$$

Substituting for *x*, *p*, and *q* we get (roughly):

> In all cases, if our colleague desires a Guinness and believes she can get one at the bar, then (all else being equal) she will go to the bar.

So who's arguing? The story might be tightened up, but the moral shines through. The explanatory force of our ordinary account depends on tacitly treating the behavior or movement as falling under general psychological laws. Otherwise, we would not have an explanation at all. What constitutes the theoretical content of folk psychology is just this framework of general laws that underpins our daily understanding of one another (see Churchland 1981, 68–69).

3 *The Trouble with Folk*

Now for the bad news. Folk psychology, it seems to some, is in various ways flawed and unsatisfactory (Churchland 1981; Stich 1983). Specific complaints about folk psychology include the following:

> (1) Folk psychology affords only a local and somewhat species-specific understanding. It flounders in the face of the young, the mad, and the alien.
>
> (2) It is stagnant and infertile, exhibiting little change, improvement, or expansion over long periods of time.
>
> (3) It shows no signs as yet of being neatly integrated with the body of science. It seems sadly disinterested in carving up nature at neurophysiologically respectable roots.

The folk, in short, just don't know their own minds. Let us look at each complaint in a little more detail.

Complaint (1) surfaces in Churchland's insistence that the substantial explanatory and predictive success of folk psychology must be set against its failure to cope with, "the nature and dynamics of mental illness, the faculty of creative imagination, ... the ground of intelligence differences between individuals, ... the nature and psychological functions of sleep, ... the miracle of memory, [and] ... the nature of the learning process itself" (Churchland 1981, 73).

Stephen Stich, in a similar vein, is worried by the failure of folk psychology to successfully explain the behavior of exotic folk and animals. Thus, he claims that we are unable adequately to characterize the content of alien or outlandish beliefs. Giving the example of someone who seems to believe he is a heap of dung, Stich notes that we are tempted to say that if it seems that someone believes *that*, then we really can't be sure *what* they believe. That is, it seems unlikely that any such alien belief can be adequately captured by any ordinary sentence of our language. Another example again developed by Stich (1983, 104) concerns the description of a dog as believing there is a squirrel up an oak tree. In one way, the ascription of such a belief seems fair enough: the dog saw the squirrel go up the oak tree and now sits at the bottom waiting for it to come back down. But in another sense it seems quite unwarranted to credit the dog with the belief that what it sees is a squirrel. (Does it also believe it sees an animal, or something that stores nuts?) Stich's point, then, is that in cases of exotic or animal beliefs, folk psychology shows signs of breaking down. It seems to urge both that someone does, and that they cannot, believe they are a heap of dung. It seems to urge that the dog does, and does not, believe there is a squirrel up a tree. Folk psychology, it seems, is just not up to taking on the really hard cases (Stich 1983, 101). And if we believe Stich and Churchland, so much the worse for folk psychology as a theory of mental life.

Moving on to complaint (2), we face the accusation that folk psychology has a history of retreat, infertility, and decadence, whereas a good theory should exhibit progress, refinement, and expansion. The thought here is that considered as a standard, speculative scientific theory, folk psychology would seem to be degenerating in the strict sense of Lakatos (1974, 91–196). A scientific theory or theory sequence is said to be degenerating if it fails over a long period to extend its early successes and to predict and explain novel phenomena. Because the theoretical substratum of folk psychology is *implicit,* it is somewhat difficult to see exactly what the point about degeneration can amount to. Churchland makes his complaint by saying, "The [folk psychology] of the Greeks is essentially the [folk psychology] we use today and we are negligibly better at explaining human behavior in its terms than was Sophocles. This is a very long period of stagnation and infertility for any theory to display. [Its] failure to develop its resources and *extend its range of successes* is therefore darkly curious and one must query the integrity of its basic categories" (Churchland 1981, 74; my emphasis). Presumably, then, the thought is that our daily explanations of each other's behavior ought ideally to be increasing in variety (new terms and phrases) and hence in detail, predictive power, and success. This would be evidence of a progressive underlying theory. One immediate comment is that something like this actually does take place. New terms and phrases are coined, and they do seem to bring increased understanding. Terms like "mauvaise foi," "Schadenfreude," and perhaps even Freudian notions of the ego and id are cases in point. (I owe these examples to Robert Griffiths.) So if there really is stagnation and infertility, it must be located at a much deeper level. And indeed, as remarked above, it is true to say that the basic framework of ascribing *beliefs* and *desires* as explanations of actions is pretty much unchanged across vast stretches of historical time and geographically distant cultures. But this level of unchanging commonality may be evidence, as we shall later see, that what we are dealing with is something rather different to a mere stagnant folk theory.

Lastly, complaint (3) folk psychology seems to show no signs of carving up nature at neurophysiologically respectable joints. Thus, Churchland's celebration of the growing synthesis of "particle physics, atomic and molecular theory, organic chemistry, evolutionary theory, biology, physiology and materialistic neuroscience" is cut prematurely short by the sad observation that folk psychology "is no part of this.... Its intentional categories stand magnificently alone, without visible prospect of reduction to that larger corpus. A successful reduction cannot be ruled out ..., but [its] explanatory impotence and long stagnation inspire little faith that its categories will find themselves neatly reflected in the framework of neuroscience" (Churchland 1981, 75). Neat reduction, it seems, is the name of the game. Commonsense ascriptions of mental states must map onto theoretical divisions within a successful scientific account of states of the head or perish. Already they are heading for the hills. But the panic, as we shall see, is somewhat premature.

Just to round off the "bad news," we might mention a general sentiment said to be shared by Stich, Churchland, and Daniel Dennett (see Stich 1983, chapter 11, note 10). This is that folk psychology is almost *bound* to prove deeply misguided.

> The very fact that [folk psychology] is a *folk* theory should make us suspicious. For in just about every other domain one can think of the ancient shepherds and cameldrivers whose speculations were woven into folk theory have a notoriously bad track record. Folk astronomy was false astronomy and not just in detail.... However wonderful and imaginative folk theorising and speculation has been it has turned out to be screamingly false in every domain in which we now have a reasonably sophisticated science. (Stich 1983, 229.)

There is surely something very wrong with this picture. Can we really imagine that our ancestors sat around a campfire and just *speculated* that human behavior would be usefully explained with ideas of belief and desire? Surely not. Some such understanding, though not verbally expressed, seems more likely to be a *prerequisite* of a highly organized society of language users than a function of their speculations. Moreover, what makes Churhland's comments *criticisms* of folk psychology, as opposed to observations about its nature? There seem to be all sorts of assumptions here about the *role* of ordinary ascriptions of mental states in our lives. Are such ascriptions really just a tool for explaining and predicting others' bodily movements? And even if in some sense it *is* such a tool, is it really trying to fulfil its purpose by tracking states of the head? Would it even be wise to try to explain behavior in such a way? If any of these pointed queries draws blood, the honor of the folk may be preserved. Instead of losing at protoscience, the folk turn out to be winning at a different game. The suspicion of a deep mismatch between the game of folk psychology and the game of scientific theorizing about states of the head has been gaining ground in recent analytic philosophy. It is worth spending a moment to reconnoiter the new terrain.

4 *Content and World*

The content of a mental state is what gets picked out by the "that" clause in constructions like: "Daredevil believes that Elektra is dead," "Mary hopes that Fermat's last theorem is true," and so on. Since the discussion of content covers questions about meaning and about mind, in such discussions philosophy of mind, philosophy of psychology, and philosophy of language all meet up, with spectacular pyrotechnic results (see, e.g., Evans 1982 and essays in Woodfield 1982 and in Pettit and McDowell 1986). The part of the display that interests us here concerns the debate over what has become known as *broad* or *world-involving* content.

There is a tendency to think of psychological states as, in essence, self-contained states of the individual subject. That is to say, of course, not that we are not located in and affected by the world, but only that our *psychological* states are not essentially determined by how the world about us really is so much as by how it strikes us as being. In other words, the intuition is that whatever doesn't in any way impinge on your conscious or unconscious awareness can't be essentially implicated in any correct specification of your mental state. On this view your mental states have the contents they do because of the way you are, irrespective of the possibly unknown facts about your surroundings.

Much recent philosophy is characterized by a snowballing crisis of faith in this seemingly impregnable doctrine. Content, according to the heretics, essentially involves the world (Pettit and McDowell 1986, 4). The crisis began on twin earth (see Putnam 1975a). The twin earth thought experiments work by varying the facts about the environment while keeping all the narrowly specifiable facts about the subject constant. Narrowly specifiable facts about a subject include the subject's neurophysiological profile and any other relevant facts specifiable without reference to the subject's actual surroundings either present or past. The upshot of such thought experiments is to suggest that some content, at least, essentially involves the world. Thus, to use the standard, well-worn example, imagine a speaker on earth who says "There is water in the lake." And imagine on twin earth a *narrow* doppelgänger (someone whose narrowly specified states are identical with the first speaker) who likewise says "There is water in the lake." Earth and twin earth are qualitatively identical except that water on earth is H_2O while water on twin earth is XYZ, a chemical difference irrelevant to all daily macroscopic water phenomena. Do the two speakers mean the same thing by their words? It has begun to seem that they cannot. For many philosophers hold that the meaning of an utterance must determine the conditions under which the utterance is *true*. But the utterances on earth and twin earth are made true or false by the presence or absence of H_2O and XYZ respectively. So *if* meaning determines truth conditions, the meaning of statements involving natural kind terms (water, gold, air, and so on) can't be fully explained simply by reference to narrowly specifiable states of the subject. And what goes thus for natural kind terms also goes (for similar reasons) for demonstratives ("that table," "the pen on the sofa," etc.) and proper names. The lesson, as Putnam would have it, is that "meanings just ain't in the head."

At this point, according to Pettit and McDowell (1986, 3), we have two options. (1) We could adopt a composite account of meaning and belief in which content depends on both an internal psychological component (common to the speakers on earth and twin Earth) *and* an external world-involving component (by hypothesis, not constant across the two earths). Or (2) we could take such cases as calling into question the very idea that the psychological is essentially inner and hence as calling into question even the idea of a purely inner and properly psychological *component* of mental states, as advocated in (1). As Pettit and McDowell (1986, 3) put it, "No doubt what is 'in the head' is causally relevant to states of mind. But must we suppose that it has any constitutive relevance to them?" Of course, we do not *have* to take the twin earth cases in either of the ways mentioned above. For one thing, they constitute an argument only *if* we antecedently accept that meaning should determine truth conditions. And even then there might be considerable room for maneuver (see, e.g., Searle 1983; Fodor 1986). In fact, I suspect that as *arguments* for content that involves the world, the twin-earth cases are red herrings. As Michael Morris has suggested in conversation, they serve more to clarify the issues than to argue for a particular view. Nonetheless, the idea that contentful states may essentially involve the world has much to recommend it (see especially the discussion of demonstratives in Evans 1982).

This, however, is not the place to attempt a very elusive argument. Instead, I propose to conduct a conditional defence of cognitive science. Even if all content turned out to radically involve the world (option (2) above), that in itself need not undermine the claim of cognitive science to be an investigation that is deeply (though *perhaps* not constitutively) relevant to the understanding of mind. In short, accepting option (2) (i.e., rejecting the idea that the psychological is essentially inner) does *not* commit us to the denial of conceptual relevance, implied in the quoted passage from Pettit and McDowell.

The notion of constitutive relevance will be amplified shortly. First, though, a word about an argument that (if it worked—which it doesn't) would make any defence of cognitive science against the bogey of broad, world-involving content look strictly unnecessary. The argument (adapted from Hornsby 1986, 110) goes like this:

> Two agents can differ in mental state *only* if they differ somehow in their behavioral dispositions.
>
> A behavioral difference (i.e., a difference in behavioral dispositions) requires some internal physical difference.
>
> So there *can't* be a difference in mental states without *some* corresponding difference of internal physical states (contrary to some readings of the twin-earth cases).

In other words, the content of mental states *has* to be narrowly determined if we are to preserve the idea that a difference in behavioral disposition (upon which mentality is said to supervene) requires a difference of inner constitution.

This argument, as Hornsby (1986, 110) points out, trades on a fluctuating understanding of "behavior." In the first premise "behavior" means "bodily movements." This is clear, since *no* state of the head can cause you to, e.g., throw a red ball or speak to Dr. Frankenstein or sit in that (demonstratively identified) chair in the real absence—despite appearances, let's presume—of a red ball, Dr. Frankenstein, or the chair, respectively. At most, a state of the head causes the bodily movements that might count, in the right externally specified circumstances, as sitting in that chair, throwing the red ball, and so on. In the second premise, however, the appropriate notion of behavior is not so clearly the narrow one. It may be (and Putnam's arguments were meant to suggest that it *must* be) that the correct ascription of contentful states to one another is tied up with the actual states of the surrounding environment. If this is so, then it would be reasonable to think that since the ascription of contentful states is meant to explain behavior, behavior *itself* should be broadly construed. Thus, there could be no behavior of picking up the red ball in the absence of a red ball (*whatever* the appearances to the subject, his bodily movements, etc.). In this sense the idea of behavior implicated in mental-state ascriptions is more demanding than the idea of mere bodily movements. In another sense, as Hornsby also points out (1986, 106–107), it might be *less* demanding, as fine-grained differences in actual bodily movement (e.g., different ways of moving our fingers to pick up the red ball) seem strictly irrelevant to the ascription of psychological states. Ascriptions of folk-psychological contents thus seem to carve reality at joints quite different to any we may expect to derive from the solipsistic study of states of the head that determine bodily movements. The conclusion (that ascriptions of folk-psychological mental states are concerned only with narrow content) is thus thrown into serious doubt. It is not clear that we can make sense of any appropriate notion of narrow content. And equivocation on the meaning of "behavior" cannot be relied upon to fill in the gaps.

The radical thesis that all ideas of content essentially involve the world thus survives the latest assault. Despite the doubts voiced earlier, I propose, as I said, to give away as much as possible and *accept* this thesis, while still denying the pessimistic "implications" for cognitive science. (This may seem a curious project, but there are independent reasons for requiring some such defence, as we shall see.)

The target posture (accept broad content and deny the conceptual or philosophical irrelevance of cognitive science) may seem uncomfortable if you accept the following argument.

> The mental states of folk psychology (belief, desire, fear, hope, etc.) are individuated by appeal to a broad, world-involving notion of content.
>
> The accounts and explanations given by cognitive science, insofar as they are formally or computationally specifiable, must in principle be independent of any semantic, world-involving considerations. They must have an internal syntactic reading that treats only of narrow, solipsistically definable states of agents. (See, e.g., Fodor 1980a.)
>
> There is every cause to believe that semantic, world-involving accounts and solipsistic, narrow accounts will not carve nature at the same joints. There will be no neatly individuated internal states (either neurophysiologically or formally specified) that map onto the mental states individuated by folk psychology.
>
> So cognitive science can't be in the business of contributing to a philosophical understanding of the nature of mental states, because the states with which it directly deals do not map satisfactorily onto our notions of mental states.

The conclusion, which is essentially that of McCulloch (1986), amounts to "a pretty fundamental rejection of the idea that there can be a scientific synopsis (of the manifest 'folk-psychological' image of ourselves and the scientific one) given that *mind* is unquestionably one of the things that must show up in it in some suitably scientific guise" (McCulloch 1986, 87–88). The question, I think, is what counts as a scientific synopsis here. Must a satisfactory synopsis involve a state-for-state correlation? Or is there some more indirect way to achieve both synopsis and conceptual relevance? I believe there is, but we must tread very carefully.

5 *Interlude*

"What a curious project!" you may be thinking. "The author proposes to attempt a defence of the significance of cognitive scientific investigations against a radical, intuitively unappealing, and inconclusively argued doctrine. And he proposes to do so not by challenging the doctrine itself, but by provisionally accepting it, and *then* turning the aggressor's blade." The reason for this is simple. With or without the broad-content theory, it looks extremely unlikely that the categories and classifications of folk psychology will reduce neatly to the categories and classifications of a scientific account of what is in the head. This is the "failing" that Churchland and (to a lesser degree) Stich ascribe to folk psychology. I embrace the mismatch, but not the pessimistic conclusion. For folk psychology may not be playing the same game as scientific psychology, despite its deliberately provocative and misleading label. So I take the following to be a very real possibility: whenever I entertain a thought, it is completely individuated by a state of my head, i.e., the content of the thought does not essentially involve the world, but there will be no projectable predicates in a scientific psychology that correspond to just that thought. By "no projectable predicate" I mean no predicate (in the scientific description) that is projectable onto other cases where we rightly say that the being is entertaining the same thought. Such other cases would include myself at different times, other humans, animals, aliens, and machines.

Regardless of broad content, I therefore join the cynics in doubting the *scientific* integrity of folk psychology as a theory of states of the head. But I demur at both the move from this observation to the conclusion that cognitive science, as a theory of states of the head, has no philosophical *relevance* to the understanding of mind (Pettit and McDowell) and the move to the conclusion that folk psychology be eliminated in favor of a scientific account of states of the head (Churchland).

What I try to develop, then, is more than just a conditional defence of cognitive science in the face of allegations of broad content. It is also a defence of cognitive science despite any mismatch between projectable states of the head and ascriptions of specific beliefs, desires, fears, etc. Relatedly, it is a defence of belief-desire talk against any failure to carve nature at internally visible joints. Coping with the broad-content worry is thus really a fringe benefit associated with a more careful accommodation of commonsense talk of the mental into a scientific framework. So, now that we know we are getting value for our money, let's move on.

6 *Some Naturalistic Reflections*

At this point I think we may be excused for indulging in a little armchair, naturalistic reflection. It seems a fair question to ask, What earthly use is the everyday practice of ascribing mental states to one another using the apparatus of folk psychology, that is, the apparatus of propositional-attitude ascription with notions of belief and desire? One answer might be that it is useful as a means of predicting and explaining other people's bodily movements by attempting to track internal states of their heads. This, we saw, is what the eliminative materialist *must* believe the practice is for.[1] Otherwise, it would hardly be to the point to criticize it for failing to carve up nature at neurophysiological joints. And if that is what the practice is for, it may be in deep water. But why should we assume that it has any such purpose? Consider an alternative picture, due in part to Andrew Woodfield.[2]

On this picture, the primary purpose of folk-psychological talk is to make intelligible *to us* the behavior of fellow agents acting in the world. In particular, it is to make their behavior intelligible and predictable just insofar as that behavior bears (or could bear) on our own needs and interests. Now let us throw in a few more small facts. The other agents whose behavior we wish to make intelligible are primarily our peers, beings with four notable traits. First, they largely share our sensitivity to the world, i.e., our senses and any innate protoconceptual apparatus. Second, they share our world. Third, they share to a large extent our own most basic interests and needs. Fourth, the biological usefulness of their thoughts, like our own, involves their tracking real states of the world, a purpose for which we may (on evolutionary grounds) assume that their thinking is well adapted. Taken together, those traits help make plain the convenience and economy of ascribing folk-psychological content. The thoughts of our peers are well adapted to the same world as our own. So given also a convergence of needs and interests, we may economically use talk of states of the world generally to pick out the salient features of the thoughts of others. The development of a tendency to make broad-content ascriptions begins at this point to seem less surprising. Broad-content ascription, we may say, is content ascription that is sensitive to the *point* of thinking, which is to track states of the world. In general, it looks as if our thinking succeeds in this. The paradox of broad-content ascription is just that when our thinking fails (when, e.g., "that chair" is entertained in the absence of any chair), we must say that the thought (or better, the thinking) failed to have the content intended. But that seems acceptable, once we see the general reasonableness of the overall enterprise.

Even if we bracket the stuff about broad-content ascription, we still have a naturalized grip on some reasons why ascribing folk-psychological content ought not to *aspire* to track neat, projectable states of the head. For the question must then arise: whose head? According to the present account, what we are interested in is a very particular kind of understanding of the bodily movements of other agents. It is an understanding of those movements just insofar as they will bear on our own needs and projects. And it is an understanding that seems to be available any time we can use talk of the world (as in the "that" clause of a propositional-attitude ascription) to help us pick out broad patterns in other agents' behavior. Take for example the sentence "John believes that Buffalo is in Scotland." This thought ascription is useful *not* because it would help us predict, say, exactly how John's feet would move were someone to tell him that his long-lost cousin is in Buffalo or even when he would set off. Rather, it is useful because it helps us to predict very general patterns of intended behavior (e.g., trying to get to Scotland), and because of the nature of our own needs and interests, these are what we want to know about. Thus, suppose I have a forged rail ticket to Scotland and I want to sell it. I am not interested in the fine-grained details of anyone's neurophysiology. All I want to know is where to find a likely sucker. If the population included Martians whose neurophysiology was quite different from our own (perhaps it involves different formal principles even), it wouldn't matter a jot, so long as they were capable of being moved to seek their long-lost cousins. Thus construed, folk psychology is *designed* to be insensitive to any differences in states of the head that do not issue in differences of quite coarse-grained behavior. It papers over the differences between individuals and even over differences between species. It does so because its purpose is provide a general framework in which gross patterns in the behavior of many other well-adapted beings may be identified and exploited. The failure of folk psychology to fix on, say, neurophysiologically well-defined states of human beings is thus a virtue, not a vice.

7 *Ascriptive Meaning Holism*

The preceding section offers what is in effect just a slightly different way of putting the fairly common observation that belief ascription (and propositional-attitude ascription in general) is *holistic*. It is a net thrown over a whole body of behavior and is used to make sense of the interesting regularities in that behavior. For this reason beliefs are ascribed in *complexes*. As one well-known meaning holist puts the point, "In saying that an agent performed a single intentional action, we attribute a *very* complex system of states and events to him" (Davidson 1973, 349). It is important, however, to be clear about exactly what meaning holism involves. In a recent attack on the doctrine Fodor summarizes it as follows: "Meaning holism is the idea that the identity—specifically, the intentional content—of a propositional attitude is determined by the totality of its epistemic liaisons" (1987, 56). An epistemic liaison of a proposition *p* is any proposition that the agent takes to be relevant to the semantic evaluation of *p*, i.e., to the determination of its truth or falsity.

Fodor rightly, I think, pours scorn on such a doctrine. For one thing, if the content of a belief *is* so determined by the totality of such liaisons, it seems unlikely that any of us very often succeeds in sharing a belief or intentional state. (See Fodor 1987, 56–57.) Fodor thus denies that the content of a belief is dependent on its epistemic liaisons. Instead, he believes that beliefs have their contents *severally*. He thus chooses to bet on a form of denotational semantics in which beliefs get their contents by brain states entering one by one into causal relations with the world. Thus, he argues that a creature "could have the concept HORSE *whether or not* it has the concept COW" and that "the thought that three is a prime number could constitute an entire mental life" (Fodor 1987, 84, 89).

Now we can begin to see what is going wrong. Fodor is assuming that the crux of the meaning holist's argument has to do with *epistemic liaisons*. In that case the point about the thought involving "3" and that involving "cow" are on a par. A more persuasive version of meaning holism focuses instead on the conditions of the *warranted ascription* of particular mental states to a being. And this in itself, at least, is not a point about epistemic liaisons as defined above. Thus, for example, we would want to know under what conditions we would be justified in ascribing to a being a grasp of the concept of three. And here it does seem plausible to insist that the concept is only ascribable to a system when it exhibits sophisticated behavior with other numbers, with mathematical functions, perhaps with the counting of external objects, and so on. If such is the behavior, we acquire in one go the warrant to ascribe a *host* of mathematical beliefs. But if the behavior is not like that, we surely forfeit the right to ascribe *any*. This requirement to *ascribe* beliefs holistically underlies, I believe, the best versions of meaning holism. Thus understood, the doctrine is surely a sensible one. How could some syntactic brain state warrant ascribing a belief about selling to a system if the system could not show by its behavior that it can also have beliefs about buying?

This kind of ascriptive holism makes perfect sense in my picture of the point of belief ascription. On that picture we ascribe beliefs by throwing a kind of interpretative net over a whole body of behavior. And the mesh of that net is guaged to our particular interest in making sense of behavior. The knots (i.e., the particular ascriptions of beliefs and desires) need not correspond to any natural, projectible divisions in whatever underlying physical or computational structures make possible the behaviors concerned (and I, for one, can't see why they are even *likely* to). When we examine connectionist models in part 2, we shall also see how a system can produce semantically systematic behavior without any internal mirroring of the semantically significant parts of the sentences we use to describe its behavior.

8 Churchland Again

We can now return to Churchland's specific criticisms of folk psychology. There were three, recall.

- The explanatory power of folk psychology is limited. The exotic, insane, and very young are left mysterious.
- It is a stagnant and sterile theory that has stayed the same for a long period of time.
- It fails to integrate neatly with neuroscience.

The last objection is now easily fielded. The failure to find a neat, boundary-preserving reduction of the categories and claims of folk psychology to projectible neuroscientific descriptions need not count as a black mark for folk theory. We may insist that the folk were not even *attempting* to theorize about projectible internal states. The concern was rather to isolate as economically as possible the salient patterns in the behavior of other agents. Since such patterns may cut across projectible descriptions in a scientific language dealing solely with internal states, it is to the credit of folk psychology that it fails to fix on descriptions with neurophysiological integrity. Nor need failure to deal with the exotic, the young, or the insane surprise or bother us. For as Stich, for example, clearly sees, the folk-psychological method exploits commonalities in environments and cognitive natures to generate an economical and appropriately located grasp of the salient patterns in the behavior of our peers. If content ascription thereby breaks down in extreme cases, so be it. That does not impugn its value as a tool in its intended domain of application. Likewise, stagnation need cause no loss of sleep. As a tool for its intended purpose, folk-psychological talk has been well shaped by the constant pressures in favor of a successful understanding of others. Such pressures may yield the form of a good solution more strongly than we often believe. The shape of the tool was forged many centuries ago. Because there are salient regularities in the environment and because the use of folk-psychological talk has a limited goal of making the behavior of others intelligible to us to just the degree necessary to plot their moves in relation to our needs and interests, it is not surprising that the hard core of such understanding (i.e., ascriptions of beliefs and desires) has remained relatively constant across temporal and geographical dimensions.

A more speculative account of the stagnation might explain the relative constancy of folk-psychological understanding by positing an innate element. On this account, belief and desire ascriptions are the mind's way of making sense of itself and others. Though strictly unnecessary to my argument, I believe this hypothesis is more reasonable than it appears at first sight. It is extensively defended in Clark 1987a, a paper that is unfortunately seriously marred by my conceiving the goals of folk psychology to be solely the tracking of neurophysiologically sound states of the head. Yet the central claim of that paper fits just as plausibly into the new treatment developed above. It is that the basic framework of a folk-psychological understanding may well be innately specified. If this is the case, conceiving of ourselves and others as moved by beliefs and desires could be as natural and inevitable as seeing the world in three dimensions. Yet in the latter case, no one talks of "folk vision" or of vision as a stagnant theory of the world. And this is despite the fact that vision fixes on some categories that physics finds irrelevant. Nor need an innate folk-psycholgocial competence surprise us. As social creatures it is vitally important that we quickly come to grips with the behavior patterns of our peers. Just as the physically mobile need to know about depth and sometimes color, so the socially mobile may need to know about beliefs and desires. For a sound psychological understanding of others must surely make an important contribution to the overall fitness of a social animal. As Nicholas Humphreys (1983) points out there will always be substantial evolutionary pressure on social animals to become more efficient natural psychologists. For in the case of such animals the other members of the group are often the single most significant factor in the animal's environment for flourishing. To take just a single example, consider a recent case study of rhesus macaque monkeys (Harcourt 1985). To prosper, these animals seem to be required to make quite sophisticated judgments concerning the motives of their peers. Very briefly, support from a high-ranking female tends to be decisive in combat situations. The likelihood of such support is increased by grooming such females. Thus, if one macaque sees another groom a high-ranking female, he must try to avoid contests with that macaque in the near future. Some knowledge of the likely behavior of others in lending and withholding support is essential to success. It does not seem unduly generous to describe the observations of grooming and predictions of future behavior as involving some primitive understanding of the motivational states of other members of the group (see Tennant 1984a, 96; Tennant and Schilcher 1984, 178; also see Smith 1984, 69). If this is right, a psychological understanding of one's fellows is as important for the success of a social animal as recognizing food or predators. No one baulks at an evolutionary account of innate competences subserving our capacities to achieve these latter goals. Why not extend this generosity to the psychological realm? If we do so, we must revise our ideas about the basis of commonsense psychological understanding, for this would not be the rambling speculations of ancient shepherds and camel drivers. Innateness, it must be admitted, is no guarantee of truth. But if the goal of folk psychology is as I have painted it, the distinction between truth and usefulness looks problematic. At any rate, ways of thinking that survive the hard knocks of the school of evolutionary testing must have something going for them. In such cases, stagnant waters run deep. (For a fuller discussion of these and related issues see Cosmides 1985; Premack and Woodruff 1978; Baron-Cohen et al. 1985; Clark 1987a; and for an opposing view see Churchland and Churchland 1978.)

With or without any innate element, in the light of all this it seems fair to reject the characterization of our commonsense understanding of mental states as a folk theory. For that characterization is tied up with the idea that such talk aims in its bumbling way to do what a good scientific theory of brain states would do better. But this is not so. A better parallel might be drawn with Hayes's 1979 conception of a naive physics. Naive physics is a body of commonsense knowledge of physical laws and concepts that helps us to get around our everyday world of macroscopic objects. Knowing (perhaps nonverbally) such concepts and relations as fluid, cause, support, above, below, and beside is vital to a mobile, manipulative being. (A leaping monkey, as Boden [1984a, 162] points out, must have some kind of grasp of distance, flexibility, support, and so on.) The most vital and basic elements of such a naive physics must be either innately specified (Boden [1984a] cites the visual-cliff experiments on newborn animals as evidence of some innate grasp of depth) or else must flow directly from the operation of probably quite specialized learning capacities trained on the available data (e.g., visual and tactile data). But the point for now is that *however* we get whatever knowledge of naive physics we have, *what* we get is an implicit understanding of just those features of the physical environment most centrally relevant to our daily projects. If the categories of a naive physics fail to match those of physical science, this is no deficit. So long as these ideas serve our daily needs, they have all the integrity they require. If it would be useful to equip a robot with a grasp of naive physics, why not also a grasp of naive psychology? In fact, the latter may be even *more* pressing. For the physical world has no conception of itself. But human agents *do* have conceptions of themselves, and they involve the commonsense ideas of mental states. Successfully living in such a community may well require robots to have a grasp of the framework used by us to conceive of our own actions. For all these reasons I object to the term "folk psychology" and prefer the more neutral "naive psychology" or "mentalistic understanding." Generally, however, I shall capitulate to the now-standard usage. With the honor of our everyday mentalistic discourse thus secure, there remains the issue of the *relation* between such discourse and work in cognitive science. Two questions loom large here. (1) Should mentalistic discourse even form the *starting point* of genuine scientific inquiry into the inner well-springs of thought? (2) How, if at all, can cognitive science claim to be contributing to a properly philosophical understanding of mind? If it at best illuminates the causal backdrop of the practice of mentalistic discourse, how can it contribute to a constitutive or otherwise philosophically interesting understanding of the nature of mental states?

The first question has an obvious answer. Cognitive science *must*, of course, rely on a folk-psychological understanding at every stage, for we need to see how the mechanisms we study are relevant (even in a merely causal sense) to our performance in various cognitive tasks. Such tasks (e.g., believing at least some of the logical implications of our other beliefs, recognizing red blocks, planning a day's work) are necessarily specified in folk-psychological terms. If the task is to model thought, we could not even know whether we had succeeded or failed without using a folk-psychological understanding to individuate the target thoughts. Scientific psychology thus ultimately answers to folk psychology in precisely the same way as physics ultimately answers to observations. There may be many layers of intervening theory, but the ultimate goal most always be to do justice to the observed phenomena (see, e.g., Van Fraassen 1980). The thought that a scientific psychology might avoid being likewise answerable to our daily understanding of mind is made plausible only by relying on the ambiguity of "behavior" mentioned earlier. If the goal of a scientific psychology were *simply* to explain bodily movements, then it could proceed independently of our folk-psychological understanding. But it is hard to see how *that* study would merit the name "psychology" (see Hornsby 1986).

Folk psychology thus defines the *goals* of cognitive science (*what* has to be explained), and it is consequently implicated in any assessment of the success or failure of cognitive science. This is because cognitive science seeks at a minimum to shed light on the causal antecedents of our capacities to behave in ways that merit description in an intentional, content-ascribing vocabulary. And we therefore depend on these semantic descriptions to assess the causal and explanatory value of the theories put forward. This, however, is not to say that the actual pictures of inner states put forward by cognitive science need *themselves* involve any essential use of terms drawn from folk psychology. Indeed, we may insist that they *not* use such terms, at least insofar as they are meant to have their usual sense (i.e., their broad, contentful, world-involving sense). The parallel with physical theory makes this clear. Observation at the normal, unaided, human level is both the *start* and the *touchstone* of physical theory. But advanced physical theories themselves do not deploy ordinary kinds of observations. Rather, they may choose to carve up the world in ways quite different to those of commonsense observation. I do not mean to push this parallel too far. My aim is simply to illustrate the possibility of a science being grounded in, and answerable to, a range of observational evidence that does not appear (and perhaps *could not* appear, in the case of broad psychological states) in the detailed accounts generated within the science in question. If cognitive science both begins with, and is answerable to, the folk-psychological concept of mind, we must reject any strong version of methodological solipsism. Cognitive science cannot, and typically does not, proceed in any kind of semantic vacuum. And this is true despite the accepted lack of fit between scientific, head-bound kinds and broad, world-involving psychological kinds.

The second question, concerning the alleged philosophical impotence of cognitive science in the study of mind, is, alas, more serious. How can cognitive science contribute to a philosophical analysis of mind?

9 *Cognitive Science and Constitutive Claims*

There is without doubt a connection between the broad-content theorist's worry that cognitive science can't illuminate the mind and the eliminative materialist's worry that folk psychology is a distorting influence on any scientific study of mind. Each party sees the same bumps and potholes in the cognitive terrain, the same deep-seated mismatch of folk-psychological kinds and narrowly specified scientific kinds. But one side concludes that the folk just don't know their own minds, while the other concludes that cognitive science can't know the folks' minds. Each side clasps mind tightly to its bosom, pitying the other side as embracing at best a distorted shadow of the real thing and deprived forever of the joys of true constitutive involvement. But this is surely overly romantic. I shall sketch a more permissive approach. First, though I must unpack the notion of constitutive relevance.

The intended contrast, as far as I can see, is between constitutive and merely causal relevance. What has constitutive relevance is somehow conceptually bound up with the subject of study (in this case, mind), whereas various factors may be *causally* relevant to thought without the tie being so tight that the very idea of thought is unable to survive their subtraction. The contrast, I suspect, is not as hard and fast as some of those who use (and often abuse) the term seem to believe. If by intellectual reflection we can see that a certain phenomenon could not occur in any physically possible world (i.e., in any world where the laws of physics apply), is this a case of constitutive or merely causal relevance? Despite such unclear cases we can, I think, make enough of the distinction to find it intelligible. It is perhaps clearest in the case of games. The rules of a game are constitutive insofar as they "create or define new forms of behaviour. The rules of football or chess, for example, ... create the very possibility of playing such games. [They constitute] an activity the existence of which is logically dependent on the rules." (Searle 1969, 34). A fact *p* is as we said constitutively relevant to a phenomenon *q* if *p* is not merely causally implicated in *q* but somehow conceptually bound up with the very possibility of *q*. Let us capture this by a principle of subtraction. Fact *p* is constitutively relevant to *q* if on conceptual grounds we can see that *q* could not survive the subtraction of *p*. Thus, to return to Searle's example, the rules of chess are not, as it were, part of the mechanics that makes chess possible; rather, they are part of what it is for a game to *be* chess. We might imagine that certain memory capacities and a universe obeying certain physical laws constitute causally necessary conditions for the possibility of actually playing chess. But there is no conceptual link between the very idea of chess and the idea of such conditions. Likewise, it may be that no being could in fact think were it made entirely of gas. But there is no *direct* conceptual link between being a thinker and not being gaseous.

The major worry can now be appreciated. It is that though causally relevant to our thinking, the scientific stories of what goes on in the head can stand in no conceptual relation to the notions of thought and meaning, and hence, from a certain rather purist viewpoint, they lack philosophical interest. This is the idea canvassed by the radical broad-content theorist depicted by Pettit and McDowell (section 4 above). But even if we accept broad content and disallow any "dual-component" approach (see McGinn 1982), it is hard to see what forces the conclusion that the scientific story lacks constitutive bite. Presumably, the thought is that any hope of a constitutive scientific account depends on finding a neat, boundary-preserving mapping between scientific kinds and mental kinds. But why should that be so?

Here is an alternative way of being constitutively relevant. Suppose that the proper ascription to one another of contentful psychological states is *doubly* broad. It is broad first because the existence of such states depends (suppose) on states of the external world. But it is broad second because the existence of such states is *conceptually* bound up with their being grounded in the *right kind* of inner causes. The picture I am proposing looks roughly like this: The correct ascription of psychological states is constitutively related to the actual and counterfactual behavior of the subject. Behavior is broad in the sense of involving not just movements but their semantic and world-involving specification, but it is also broad in that it involves not just any old internal cause of the movements but a cause of a certain formally specifiable kind (see chapters 5 to 9). Just as we might revise our description of the behavior on receipt of new data about the being's relation to the world, so too we might revise it on receipt of new data concerning the inner cause of the bodily movements involved. To give the standard example,[3] suppose someone discovered that his neighbor's movements had all been caused by a giant look-up-tree program with precise descriptions of outputs in the form "If (input) then (output)." In this case (astronomically unlikely and maybe even physically impossible) the descriptions of the neighbor as truly acting are in fact unwarranted, or defeasibly warranted and now defeated. In the absence of the right, doubly broadly specified behavior, our earlier ascriptions of psychological states and contents are likewise defeated.

There are various complications here. For example, it may be that although the discovery of a certain computational substructure gives us warrant to *withdraw* ascriptions of mentality, there is no substructure whose presence is *necessary* for the ascription of mentality. This is a very real possibility, and it leaves the constitutive status of the substructural stories uncertain. It seems as if there is logical space between fully constitutive features (meeting the subtraction principle) and merely causal supports. That space is filled by, for example, cases in which we have a *set* of features (or types of computational substructures) such that

(1) being a thinker requires deployment of *some* member of that set, and

(2) a conceptually visible relation obtains between *each* individual member of the set and the warranted ascription of contentful thought (e.g., the substructures can each be seen to support the flexible actual and counterfactual behavior that warrants the use of a mentalistic vocabulary), but

(3) there is no further formal or scientific unity to the set of structures picked out in (2) (i.e., no metalevel formal or scientific description capable of meeting the demands of the subtraction principle).

My own suspicion is that the boundary between the constitutive and the causal is too sharply drawn and that genuine conceptual interest may accrue to all kinds of cases that fail to pick out relations as strong as constitutivity.

This issue of constitutivity raises the somewhat thorny problem of philosophical relevance. Under what conditions would knowledge of the in-the-head, computational substructure of thought count as a contribution to a properly philosophical understanding of mind? In point of fact, I feel a strong resistance to such a form of question. If one goal of philosophical reflection is to achieve an integrated picture of the world and our place in it as knowers, studies of the computational backdrop of our knowing have as much potential to reveal something of the nature (and possible limitations) of human thought as any other research I can think of. And the days of hard and fast disciplinary boundaries, often dictated more by administrative convenience than academic concerns, are receding fast, thankfully. My own experience in the highly interdisciplinary School of Cognitive Sciences at Sussex University is that this blurring of disciplinary boundaries is a necessary step along the road to solving many of the major problems that the individual disciplines (e.g., philosophy, psychology, linguistics, artificial intelligence) once wanted to call their own.

Nonetheless, there is a certain point to pursuing the kinds of questions just raised under the banner of constitutive relevance. *Some* claim is clearly to be made concerning the relation between various kinds of computational substructures and the nature of human thought. But what, precisely, is the intended relation? Until we are clear about this (or if it is too early to decide, clear at least about the possible options), we cannot fulfill our goal of suggesting just how these explanatory projects might fit into an overall picture of the nature of mind. Moreover, unless we are clear about what claims are or are not being made, we will have no idea what kind of evidence would support or refute them. In pursuing questions about constitutive relations and the like in subsequent chapters, it is thus not my intention to suggest that nothing short of full constitutive relevance counts as philosophically interesting. Such a view—although common enough among philosophers—depends on a much crisper conception of the boundaries of both constitutivity and philosophical interest than any I wish to endorse.

One central feature of the position I here advocate remains to be spelled out. It is that the relation between the correct type of inner story and the required behaviors need not itself be specified on any one-to-one or neat boundary-preserving basis. Rather, we may imagine specifying in some formal way (deferred until chapters 5 to 9) a *kind* of internal structure conceptually related to the very possibility of the rich, flexible actual and counterfactual behaviors required for the ascription of mental states. An internal structure is thus deeply implicated in rich flexible behavior, which warrants simultaneously ascribing a whole host of mental states to the subject. But to repeat a point that it is almost impossible to overstress, this in no way requires or suggests any neat boundary-preserving mapping between each of the holistically ascribed mental states and scientific stories about the inner causes of the bodily movements involved. In short, I reject the picture of some neat, boundary-preserving mapping between commonsense mental states $M_1, \ldots, M_n$ and narrowly specified scientific states $S_1, \ldots, S_n$. And instead, I adopt this picture:

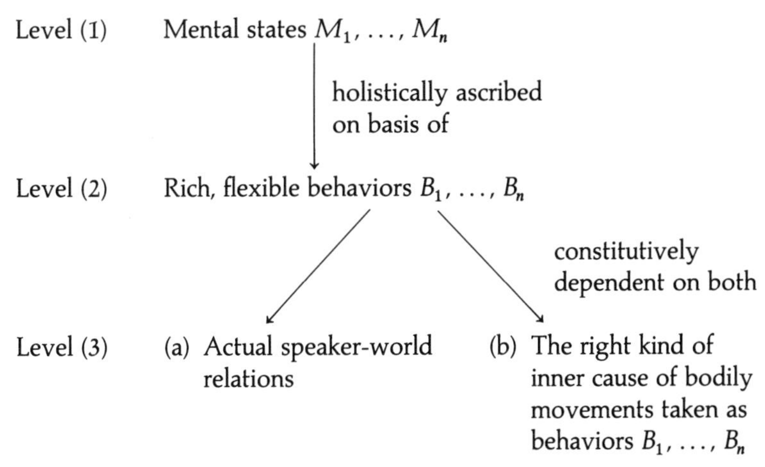

Note that the relation between levels (1) and (2) is holistic. We are warranted in ascribing groups of mental states on the basis of overall behavior. The relation between (3b) and (1) is far from a neat, boundary-respecting isomorphism. Such isomorphism is sabotaged by both the role of (3a) and the holistic nature of the relation between (1) and (2).

10 Functionalism without Folk

How does all this relate to our earlier discussions of functionalism and classical cognitivism? Think of the way extra nails relate to a coffin lid. The classical cognitivist, recall, was committed (in practice at least) to a specific architectural assertion, namely, that you could instantiate a thinking system by ensuring (at least for the in-the-head component) that it engaged in the appropriate manipulation of standard symbolic atoms. But if my earlier conjectures are at all on the mark, the attempt to model (with a view to instantiation) the inner, scientifically investigable causes of our behavior (or better, movements) by putting versions of folk-psychological descriptions *into the machine* begins to look very peculiar and unsound. It is as if we were attempting to create a human thinker by putting sentences specifying various contentful states into her head. This now looks precisely backward. On my picture, we need to specify inner states capable of causing rich, flexible behavior, which itself determines (without boundary-preserving mappings) the correctness of folk-psychological descriptions. Putting the mind's possibly broad descriptions of the states of agents acting in the world back into the head and expecting thereby to create mentality is slightly bizarre. In this respect, the eliminative materialist seems to have been right; cognitive science shouldn't seek to model internal states on ordinary contentful talk. For such talk is not able (nor, I would add, intended) to be sensitive to the relevant internal causes, yet it *is* sensitive to irrelevant external states of affairs. The contrast is between putting tokens of ordinary contentful talk back into the head (classical cognitivism) and seeking an account of how what *is* in the head enables the holistic ascription of such contents to the subject in the setting of the external world. In sum, the less plausible we find folk-psychological ideas as a scientific theory of the internal causes of behaviour, the less plausible the classical cognitivist program should seem, since it relies heavily on that level of description. Where goes the classical cognitivist, there goes the standard functionalist also. For standard functionalism (*not* the microfunctionalism described in chapter 2, section 6) is committed to filling in the following schemas for each individual mental state.

> Mental state p is any state of a system that takes x, y, z (environmental effects) as inputs, gives l, m, n, (bodily movements, vocalizations, etc.) as outputs, and fulfils internal state transitions g, h, i.

The difficulties start in the specification of g, h, i, the internal state transitions. For internal state transitions specify relations between mental states, folk-psychologically identified. The folk-psychological specifications act as placeholders to be filled in with appropriate, syntactically specified scientific kinds in due course. But this, of course, is simply to bet on the neat, boundary-preserving relation between folk psychological kinds and scientific kinds, which I have been at pains to doubt for the last umpteen pages. That kind of functionalism—the kind that treats folk-psychological descriptions as apt placeholders for scientifically well-motivated states of the head—rightly deserves most of the scorn so freely heaped on it by the eliminative materialist.

In conclusion, if I'm even halfway right, the folk *do* know their own minds. But they do so in a way sensitive to the pressures of a certain explanatory niche, the niche of our everyday understanding of patterns in behavior. The pressures on a computational theory of brain activity are quite different. Such a theory is about as likely to share the form of a folk-psychological picture of mind as a land-bound herbivore is to share the form of a sky-diving predator.

Chapter 4
Biological Constraints

1 Natural-Born Thinkers

Cognitive science is, in practice, a highly design-oriented investigation into the nature of mental processes. As we saw earlier (chapter 1, section 4), a popular methodology goes something like this. First, isolate an interesting human achievement, story understanding, say. Second, find the best way you can of getting a conventional Von Neumann computer to simulate some allegedly central chunk of the input-output profile associated with human performance of the task. Finally, hope that the program so devised will be at least a clue to the form of a good psychological theory of how human beings in fact manage to perform the task in question. This classical-cognitivist strategy, I shall now argue, is as biologically implausible as it is philosophically unsatisfactory. No serious study of mind (including philosophical ones) can, I believe, be conducted in the kind of biological vacuum to which cognitive scientists have become accustomed. In this respect at least, Pylyshyn's admiration of the post-Turing idea of a study of cognitive activity "fully abstracted in principle from both biological and phenomenonological foundations" (see chapter 1, section 2) strikes me as misplaced. Constraints that apply to naturally evolved intelligent systems are relevant to *any* attempt to model or understand the nature of human thought. This chapter simply plots some of these constraints and shows how they fuel the fires already lit underneath classical cognitivism. I treat the constraints in ascending order of importance for cognitive science.

2 Most Likely to Succeed

Remember those American high-school yearbooks in which a few characters were singled out as possessing the kinds of traits that indicate future success? In the evolutionary yearbook the star qualities are revealing and somewhat removed from the capacities most easily modeled by researchers in artificial intelligence. A preliminary list of such qualities might include real-time sensory processing, integration of various input and output modalities, capacity to cope with degenerate and inconsistent data, and flexible deployment of available cognitive resources. Such capacities subserve the goal of the organism's success in a fast-moving, competitive environment. In the service of such a goal, constant accuracy may have to be traded for speed.

Thus, to take a central case, it is a fair bet that naturally intelligent systems must cope with degenerate and even inconsistent data without crashing. If we are told both that someone is fat and that she is thin, this ought not to preclude our identifying her if we *also* know that she is wearing an orange, polka-dot suit. Having good grounds to assert *p* and good grounds to assert not *p* ought not to encourage us to allow ourselves then to infer anything at all, given the contradiction. To survive and prosper, a system must be able to cope with conflicting and inadequate data and do so without simply ceasing to act entirely. When the tiger leaps, do something, anything; don't just stand there. Ideally, systems should be able and willing to generate intelligent guesses even on the all-too-common basis of inadequate or inconsistent data. If a speech-recognition system can't hear a particular phoneme, it should guess on the basis of the phonemes it can hear and the words they could make up. The need to cope with incomplete or inconsistent data is sometimes called the requirement of *graceful degradation*. Performance should gradually become less satisfactory as available data decreases; it should not suddenly cease. Systems, in short, must be robust enough to survive in an informationally hostile environment.

Robustness of a somewhat different kind is needed to withstand the physical battering of violence, old age, and entropy. Our storage and deployment of information ought not to be overly susceptible to the loss of a few brain cells or a bang on the head. This requirement too is known as one of graceful degradation, except this time the partial lack of data is due not to external circumstances but to damage sustained by our encoding, storing, and retrieving mechanisms. Apart from being robust, the natural-born thinker should also be flexible. Some organisms choose to compete in information processing warfare; humans have and clams haven't. Dennett (1984a, 37–38) notes that this is the evolutionary choice between a Maginot line (an immobile, armoured approach) and guerilla warfare (a mobile, intelligent approach). The latter choice results in a cognitive arms race. If you do choose to compete in a cognitive arms race, one vital asset is the capacity to deploy as much of your informational armory in as wide a range of situations as possible. In particular, it should be possible to do something, even when faced with a new and unexpected situation. This, as we shall later see, may mean trying to avoid the rigid, task-bound storage of information.

Finally, I should note the pervasive demands of cognitive integration (or perhaps protocognitive integration) and real-time sensory processing. A successful organism of the mobile, guerilla-warrior class must receive and integrate data from many sensory modalities, achieve the task in real time, and control any appropriate sensorimotor activity (see, e.g., Walker 1983, 188–209). This may prove to be a surprisingly strong constraint on the kind of computational architecture required (see part 2).

In sum, the mobile organism most likely to succeed is one that is willing and able to act quickly on messy and even inconsistent data, is able to perform sensory-processing tasks in real time, is preferably able to integrate the data received through various modalities and deploy it flexibly in new situations, and is generally an all-round biological achiever. This multifaceted profile contrasts quite starkly with the kinds of systems often studied in artificial intelligence. Most of these would not last five minutes in the real world. Like overprotected children, they have been freed to develop an impressive performance along some very limited dimension (e.g., chess playing) by never having to cope with the most rudimentary tasks that any natural chess-playing organism will have coped with. This strategy of investigating intelligence in what I shall call *vertically limited microworlds* is, I suspect, a major cause of the failure of AI to contribute as much to an understanding of human psychological processes as we might have hoped. Vertically limited microworlds take you right up to something close to human performance in some highly limited and evolutionarily very recent intellectual domain. *Horizontally limited microworlds*, by contrast, would leave you well below the level of human performance right across the board but would tackle many of the kinds of tasks forced on the evolving creature at an early stage. Such horizontally limited microworlds would be, in effect, the cognitive domains of animals much lower down the phylogenetic tree than us. Flexible, robust, multipurpose, but somewhat primitive systems will (I suspect) teach us more about human psychology than inflexible, rigid simulacra of fragments of high-level human cognitive achievements. (A current example of such work is Schreter and Maurer's [1986] project on sensorimotor spatial learning in connectionist artificial organisms.) This point is further developed in section 4 below.

3 *Thrift, the 007 Principle*

Sponges feed by filtering water. Successful feeding thus requires that water pass through the sponge. To this end these small beings have developed flagella that are capable of pumping water at a rate of one bodily volume every five seconds. This much was known as long ago as 1864. Until quite recently it was assumed that this pumping action accounted for all the water that the sponge processed. Evolution, however, shows herself to be a thrifty mistress. It turns out that sponges *also* exploit the structure of their natural environment to reduce the amount of pumping required (Vogel 1981). The discovery that sponges use ambient water currents to aid their feeding was made only in the last decade. And yet, as Vogel points out:

> The structure of sponges is most exquisitely adapted to take advantage of such currents, with clear functions attaching to a number of previously functionless features. Dynamic pressure on the incurrent openings facing upstream, valves closing incurrent pores lateral and downstream, and suction from the large distal or apical excurrent openings combine to gain advantage from even relatively slow currents. And numerous observations suggest that sponges usually prefer moving water. Why did so much time elapse before someone made a crude model of a sponge, placed it in a current and watched a stream of dye pass through it? (1981, 190)

Vogel's question is important. Why *was* such an obvious and simple adaptation overlooked? The reason, he suggests, is that biologists have tended to seek narrowly biological accounts, ignoring the role of various physical and environmental constraints and opportunities. They have, in effect, treated the organism as if it could be understood independently of an understanding of its immediate physical world. Vogel believes a diametrically opposed strategy is required. He urges a thorough investigation of all the simple physical and environmental factors in advance of seeking any more narrowly biological account. He thus urges, "Do not develop explanations requiring expenditure of metabolic energy (e.g. the full-pumping hypothesis for the sponge) until simple physical effects (e.g. the use of ambient currents) are ruled out" (Vogel 1981, 182). Vogel gives a number of other examples involving prairie dogs, turret spiders, and mimosa trees.

It is the general lesson that should interest us here. As I see it, the lesson is this: if evolution can economize by exploiting the structure of the physical environment to aid an animal's processing, then it is very likely to do so. And processing here refers as much to information processing as to food processing. Extending Vogel's observations into the cognitive domain, we get what I shall dub the 007 principle. Here it is.

> *The 007 principle.* In general, evolved creatures will neither store nor process information in costly ways when they can use the structure of the environment and their operations upon it as a convenient stand-in for the information-processing operations concerned. That is, know only as much as you need to know to get the job done.

Something like the 007 principle is recognized in some recent work in developmental psychology. Rutkowska (1984) thus argues that a proper understanding of the nature of infant cognition requires rejecting the solipsistic strategies of formulating models of mind without attending to the way mind is embedded in an information-rich world. It is her view that computational models of infant capacities must be broad enough to include use of external structures as essential elements of computation. As she puts it, "The notion of computation as rule-governed structure manipulation must be taken to include environmental as well as intra-subject structures" (Rutkowska 1984, 1). This extension is vitally important. Various studies in developmental psychology support the need for such an approach. In the very simplest case (I investigate a much more complex and interesting one in chapter 7) when seeking an ingredient for baking a cake, a child does not need to remember exactly where the ingredient is located in a store. Instead, the child may simply go to the right shelf and there seek what it needs. In such cases the external world stands in for a highly detailed memory store. (This example, originally from Cole, Hood, and McDermott 1978, is cited in Rutkowska 1986, 88.)

Cognitive ethologists are likewise quick to recognize the ways in which animals have developed to augment their limited intellectual capacities with the wily use of environmental structures. Clark's nutcracker, for example, buries seeds near logs to facilitate future rediscovery.

Nor, of course, is the strategy of exploiting environmental structures to aid cognition confined to children and lower animals. Consider the practice of solving a jigsaw puzzle. This activity combines purely internal cognitations (e.g., "This section has half a bird on it, so I need a piece with a wing and a piece with a foot") with physical operations on a real object (the actual puzzle). These physical operations are essential to our problem-solving activity; we seldom represent the shape of a piece to ourselves well enough to know for sure that it will fit in advance of trying it in a plausible location. The physical operation of trying the piece out may result in a fit, in which case the ordinary, nonextended cognitive process again commences ("What will the next piece have to look like, given the shape and pictorial content of this piece?"). Solving a jigsaw puzzle (or at least human solving of jigsaw puzzles) cannot be explained purely by appeal to a set of *internal* processes interpreted as working out the steps of the solution. Rather, the internal processes must tie in with real operations on the world for testing hypotheses and generating new states of information. Imagine trying to devise a model of human capacities to solve jigsaw puzzles that took no account of our ability to manipulate the real puzzle. Any such model would achieve its goal only by constructing a complete internal representation of the shape of each piece. This may do the job, but it would hardly be a model of *human* cognition in the domain. It would instead be an example of what Dennett (1984b) calls a "cognitive wheel," an elegant but unnatural solution to a problem of natural design.

A final point to stress in this context (one that harks back to some of Dreyfus's criticisms of AI outlined in chapter 2) is that the external structures an intelligent system may exploit include both other agents and its own body. The potential uses of other agents are, I suppose, obvious enough. Two heads are indeed often better than one. Since it is so phenomenonologically immediate, the use of one's own body is easily overlooked. Jim Nevins, a researcher into computer-controlled assembly, cites a nice example (reported in Michie and Johnston 1984). One solution to the problem of how to get a computer-controlled machine to assemble tight-fitting components is to design a vast series of feedback loops that tell the computer when it has failed to find a fit and get it to try again in a slightly different way. The natural solution, however, is to mount the assembler arms in a way that allows them to give along two spatial axes. Once this is done, the parts simply slide into place "just as if millions of tiny feedback adjustments to a rigid system were being continuously computed" (Michie and Johnston 1984, 95).

A proponent of the ecological movement in psychology once wrote, "Ask not what's inside your head but rather what your head's inside of" (Mace 1977, quoted in Michaels and Carello 1981). The positive advice, at least, seems reasonable enough. Evolution wants cheap and efficient solutions to the problems of survival in a real, richly structured external environment. It should come as no surprise that we succeed in exploiting that structure to our own ends. Just as the sponge augments its pumping action by ambient currents, intelligent systems may augment their information-processing power by devious use of external structures. Unlike the ecological psychologists, we can ill afford to ignore what goes on inside the head. But we had better not ignore what's going on outside it either.

The moral, then, is to be suspicious of the heuristic device of studying intelligent systems independently of the complex structure of their natural environment. To do so is to risk putting into the modeled head what nature leaves to the world. Classical cognitivism, we shall later see, may be guilty of just this cardinal sin.

4 *Gradualistic Holism and the Historical Snowball*

A very powerful principle pervades the natural order. It is often cited, though, as far as I know, it is unbaptised. The principle was explicitly stated by H. Simon (1962) and has since been mentioned by a great many writers. The principle states that the evolution of a complex whole will generally depend on its being built out of a combination of parts, each of which has itself evolved as a whole stable unit. The recursive application of such a procedure enables us to account for the evolution of complex wholes without the explosive increase in improbability that would dog any claim that such a complex evolved in a single step. For example, the one-step evolution of such a structure as depicted in figure 4.1 is much less likely than the evolution of such a structure if there are stable intermediate forms, as illustrated in figure 4.2. The example is deliberately simplistic. But the

Figure 4.1
Complex structure with no simpler forms

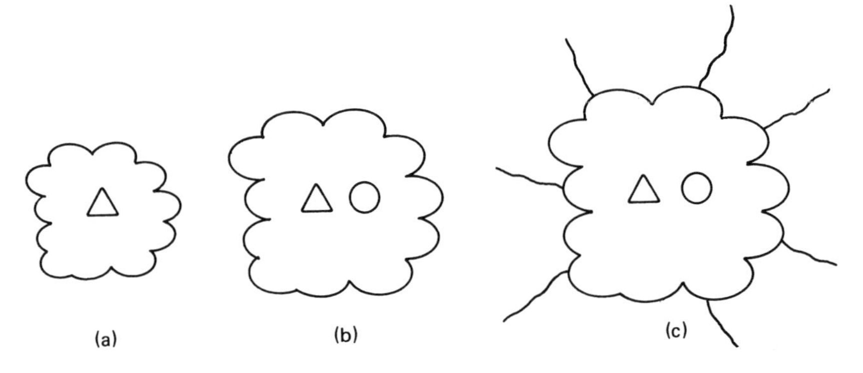

Figure 4.2
Complex structure with stable intermediate forms

principle is genuine and powerful. It covers both the combination of existing units at a given time and the succession of structures over a period of time. In the latter incarnation it surfaces as the evolutionary demand for *gradual* change that results in the success of the whole organism. Thus, suppose we imagine the structure (c) in figure 4.2 to represent an evolved being. The evolution of (c) is possible only because the previous structures (a) and (b) are themselves successful adaptations able to survive and reproduce. As Dawkins (1986, 94) recently put it, it is essential that an evolutionary trajectory should avoid passing through any disadvantageous intermediate states. An intermediate form that is maladaptive may well be a stage in the hypothetical evolution of a very well adapted being. But nature is not impressed by such farsightedness. Nature countenances only intermediate forms that prosper in the short term, even if the performance of later models is compromized as a result.

Such, in the abstract, is the general principle that I shall dub "gradualistic holism." According to gradualistic holism the evolution of a complex product is practically possible only insofar as that product is the last (or current) link in a chain of structures satisfying two conditions. First, at each stage the chain must involve only a small change in structure from earlier stages (gradualism). Second, each such stage must yield a new structure that is itself a viable whole (holism). When applied to complex biological organisms, this amounts to the requirement that the complex product emerge as the outcome of a series of small structural alterations, each of which yielded a whole organism capable of surviving and prospering in its own niche.

Biological theory countenances three ways in which the requirements of gradualistic holism may be met. These ways represent accepted solutions to the problem of *coadaptation,* i.e., the problem of accounting for the evolution of complex organisms whose parts seem made for each other yet whose simultaneous one-step evolution would require a ridiculously improbable coincidence of mutations. As outlined, e.g., in Ridley 1985, 35–41 the three solutions to the problem of coadaptation are

(1) piecemeal evolution with constant function,
(2) piecemeal structural evolution with functional change, and
(3) symbiosis.

It is worth examining these in a little detail, as they represent important alternatives with rather different implications (as we shall later see) for the cognitive domain.

The clearest example of (1) is the evolution of the eye. This story has been told often enough, and I doubt it is necessary to repeat it again (see, e.g., Ridley 1985, 36; Dawkins 1986, 80–83; Clark 1986, 54). The point is that we can describe a series of seeing organs such that the series progresses from simple collections of light sensitive cells, through pinhole camera eyes, to eyes with lenses, and thence to the eyes of vertebrates like ourselves. Each step in the series can be obtained by small structural alterations to its predecessor, thus meeting the requirement of gradualism. And each step confers some useful increment in visual acuity, thus meeting the requirement of holism. Moreover, we can find instances of each intermediate stage in living species, which is tantamount to empirical proof that the organisms have met the holistic requirement. Each link in the chain leading to the vertebrate eye was itself an organ used for seeing.

Piecemeal evolution with change of function is also possible, as indicated in option (2). The standard example here is the evolution of flight, again discussed by Ridley (1985). Feathers, it is said, may first have evolved to perform the function of thermoregulation by effectively covering the bird and trapping a cushion of air between body and feather. A bird that had already evolved a basic wing structure for simple gliding might later find that the feathers could aid flight. When an organ that initially served one purpose proves capable of subserving a subsequent and different function, we speak of its being "preadapted" for the later function. This is a bit misleading, since it suggests some foreknowledge of the subsequent use, which is precisely what we set out to deny. For consistency I shall stick with the received terminology. Feathers, then, may be described as adapted for thermoregulation and preadapted for flight. When there is such a change in function, the organ may or may not continue simultaneously to perform its original function. Feathers continue to help thermoregulation even when used for flight. According to one interpretation, our lungs evolved as breathing devices only thanks to a preadaptation in the form of the swim bladders of fish. Swim bladders are sacs of air that help a fish move in water. On one account, because our lungs evolved out of a change of function of the swim bladder, we are susceptible to ailments like pleurisy and emphysema. As Lieberman (1984, 22) puts it: "Swim bladders are logically designed devices for swimming—they constitute a Rube Goldberg system for breathing." (Rube Goldberg, I am told, is the American Heath Robinson.) Here, then, we have a case in which a series of gradual structural changes includes a large and significant change in function (from swimming aid to breathing device) and the end product (the breathing device) can be seen as a kludge if it is considered as a device designed for its current role. "Kludge" is a term used in engineering and computer science to describe something that, from a pure (i.e., ahistorical), design-oriented viewpoint, looks messy and inefficient. But it gets the job done. And it may even count as an elegant solution once all the constraints (e.g., the available skills and resources) are taken into account.

The third and final means by which biological systems may meet the demands of gradualistic holism is symbiosis. In symbiosis the various parts of a subsequent unit evolve separately and are put together at a later date. The classic example this time is the evolution of the eukaryotic cell (Ridley 1985, 38–39; Jacob 1977, 1164). A eukaryotic cell has a nucleus and contains such organelles as mitochondria and chloroplasts. It is much more complex than the prokaryotic cell, which is little more than a sac of genetic material. One account of the evolution of the complex eukaryotic cell depicts the mitochondria and chloroplasts (which now convert food into energy for the cell) as once-independent organisms that later formed a mutually desirable alliance with the host cell. The terms of the alliance need not concern us. Very roughly, the host cell is provided with energy, and the organelles with food and raw materials. Such alliances allow for great increases in complexity achieved very rapidly by the association of existing systems. (Something similar may occur in human thought when we see how to unite ideas developed in separate domains into a single overarching structure.) Symbiosis thus meets the requirements of gradualistic holism with a *contemporaneous* set of independent viable structures (thus meeting in an unusual way the demand of holism) and with a small structural alteration (as demanded by gradualism) that supports their amalgamation into a new whole.

Complex biological systems, then, have evolved subject to the constraints of gradualistic holism. This mode of evolution suggests the possibility of a snowballing effect with worrying implications for cognitive science. The snowballing effect is summed up in an informal principle formulated by Jacob, a cell geneticist. It is just this: "Simpler objects are more dependent on (physical) constraints than on history. As complexity increases, history plays the greater part" (Jacob 1977, 1163). The idea is simple, and follows immediately from our earlier observations. Gradualism requires that each structural step in the evolutionary process involve only a small adjustment to the previous state. Jacob compares evolution to a tinkerer who must use whatever is immediately at his disposal to achieve some goal. This case contrasts with that of an engineer who, within certain limits, decides on the appropriate materials and design, gathers them, and then carries out the project. (Levis-Strauss [1962] explored a similar analogy involving the notion of bricolage. See also Draper 1986.)

The point, then, is that what the tinkerer produces is heavily dependent on his historical situation in a way in which the engineer's product is not. Two good engineers will often arrive independently at a similar design that "approaches the level of perfection made possible by the technology of the time" (Jacob 1977, 1163), whereas two tinkerers attempting to use only the materials they happen to have to hand will be unlikely to chance on the same solution. If evolution proceeds as a tinkerer, each step in the evolutionary chain exploits a net historical opportunity whose nature is determined by whatever materials happen to be available to adapt to a new requirement. Chance and local factors will play some role at every stage along the way. Since every later development occurs in a space determined by the existing solutions (and materials), it is easy to see that there will be a snowball effect. Every idiosyncrasy and arbitrariness at stage s_1 forms the historical setting for a tinkering solution to a new problem at s_2. As complexity increases and s_1 gives way to s_n, the solutions will come more and more to depend on the particular history of the species. This may be one reason why some evolutionary theorists (e.g., Hull 1984) prefer to regard a species as a historical individual picked out by the particular circumstances of its birth, upbringing, and culture, rather than as an instance of a general, natural kind.

This historical snowballing effect, combined with the need to achieve some workable total system at each modification (holism), often makes natural solutions rather opaque from a design-oriented perspective. We have already seen one example in the evolution of a breathing device from a swim bladder. If we set out to build a breathing device from scratch, we might, it seems, do a better job of it.

Two further examples should bring home the lesson. These examples, beautifully described by Dawkins (1986, 92–95), concern the human eye and the eye of the flatfish. The human eye, it seems, incorporates a very strange piece of design. The light sensitive photoreceptor cells face *away* from the light and are connected to the optic nerve by wires that face in the direction of the light and pass over the surface of the eye before disappearing through a hole in the retina on their way to the brain. This is an odd and seemingly clumsy piece of design. It is not shared by all the eyes that nature has evolved. An explanation of why vertebrate eyes are wired as they are might be that the combination of some earlier historical situation with the need to achieve an immediate working solution to some problem of sight forced the design choice. The wiring of the eye, then, has every appearance of being a kludge, a solution dictated by available materials and short term expediency. As a piece of engineering it may be neither elegant nor optimal. But as a piece of tinkering, it worked.

Dawkins's second example concerns bony flatfish, e.g., plaice, sole, and halibut. All flatfish hug the sea floor. Some flatfish, like skates and rays, are flattened elegantly along a horizontal axis. Not so the bony flatfish. It is flattened along a vertical axis and hugs the sea floor by lying on its side. This rather ad hoc solution to whatever problem forced these fish to take to the sea bed must have raised a certain difficulty. For one eye would then be facing the bottom and would be of little use. The obvious tinkerer's solution is to gently twist the eye round to the other side. Overall, it is a rather messy solution that clearly shows evolution favoring quick, cheap, short-term solutions (to hug the sea bed, lie on your side) even though they may give rise to subsequent difficulties (the idle eye), which then get solved only by further tinkering. This process, incidentally, is recapitulated in the development of the bony flatfish, the bony flatfish starts life as a symmetrical surface-swimming fish and subsequently undergoes a distortion of the skull as one eye moves over the head to the other side of the fish, which then settles on the sea bed to live out its days. As Dawkins (1986, 92) puts it: "The whole skull of a bony flatfish retains the twisted and distorted evidence of its origins. Its very imperfection is powerful testimony of its ancient history, a history of step by step change rather than of deliberate design. No sensible designer would have conceived such a monstrosity if given a free hand to create a flatfish on a clean drawing board." It seems, then, that to understand the appearance of the bony flatfish, we must treat the species as a historical individual whose current state is the product of a particular series of accidents, problems, and short-term solutions to these problems.

The underlying explanatory principle of gradualistic holism and the snowball effect it induces suggest that we should treat *all* evolved complex systems similarly. If this is so, the implications for what is fundamentally a *design-oriented* cognitive science may be profound. For why suppose that cognitive adaptations are exempt from the same constraints? To put the point starkly, why suppose that our means of, say, playing chess is not fundamentally informed by the natural constraint of building a chess-playing capacity out of cognitive components designed for spotting predators? The mind, as far as I can see, is as good a place to find a kludge as the lung or the eye.

An analogy may help to clarify the idea. Imagine that you devise a range of software for a small business. As the business expands you are able to change small bits of the software in fairly circumscribed ways. What you can *never* do, even when the firm becomes a multinational corporation, is rewrite the software package from scratch. You end up trying to run the equivalent of a small country with a deviously deployed software package that still bears the hallmarks of the corner grocery shop for which it was originally devised. That is the way I see the cognitive kludge. Someone sitting down with a clean slate to devise software for a multinational would doubtless end up with a very different package, one that might boost profits quite considerably. But evolution builds its new cognizers out of old parts and seeks results with minimal alterations. Hence the kludge.

The third and final moral, then, is that a computationally oriented investigation into the principles of human psychology had better attend to the kinds of capacities required for success at a fairly evolutionarily basic level, since the solutions developed here are likely to impose very strong constraints on the later solutions to higher-level problems. The cognitive scientist thus needs at least a broad appreciation of what we may term the *functional phylogeny of mind*. Such a functional phylogeny would comprise a rough ordering of the problems facing an evolving organism competing in a cognitive arms race. The kinds of capacity that such a phylogeny would pick out as primitive, central, and hence suitable objects of ordinary design-oriented investigation might include:

Locomotor and manipulative skills

• sensorimotor coordination for walking, climbing, running, using tools, etc.

• kinesthetic and proprioperceptive awareness (knowledge of one's own movement in space and of the relative locations of one's own bodily parts)

Object-oriented skills

• recognition of objects as enduring in space and time

• recognition of the value of objects (e.g., to eat, to play with, to fear, etc.)

Spatial skills

• navigational skills

• spatial memory

• path recognition (e.g., spotting a way through dense undergrowth)

Perceptual skills

• various perceptual systems

• cross-modal abstraction and integration

General cognitive skills

• analogical reasoning, learning from experience

• selective attention

• emergency interrupt systems (e.g., "Stop feeding. There's a lion coming!")

• memory and anticipation

• modeling oneself and the environment

• curiosity, actively seeking out relations of cause and effect

• playfulness

Social skills

• recognition of others' rank or standing

• prediction and control of others' behavior by psychological modeling of the other

(The above list is provisional and incomplete. It is drawn in part from some section headings in a recent work on animal cognition [Walker 1983, 189–235].)

The list is odd in that some of the items (like emotional response and curiosity) seem either too biological or too complex for current work to address. Yet these, I suggest, are the building blocks of human cognition. It seems psychologically ill-advised to seek to model, say, natural language understanding without attending to such issues. Paradoxically, then, the protocognitive capacities we share with lower animals, and not the distinctive human achievements such as chess-playing or story understanding, afford, I believe, the best opportunities for design-oriented insight into human psychology. This is not to say, of course, that all design-oriented investigation into higher-level skills is unnecessary, merely that it is insufficient. Indeed, a design-oriented approach to the higher-level achievements may be necessary if we are to understand the nature of the task that evolution may choose to perform in some more devious way. The general point I have been stressing is just that understanding the natural solution to an information-processing task may require attending at least to the following set of biological motivated constraints:

- High value must be placed on robustness and real-time sensory processing (section 2).
- The environment should not be simplified to a point where it no longer matters. Instead, it should be exploited to augment internal processing operations (section 3).
- The general computational form of solutions to evolutionarily recent information-processing demands should be sensitive to a requirement of continuity with the solutions to more basic problems (section 4).

5 *The Methodology of MIND*

The methodology of classical cognitivism involved a valiant attempt to illuminate the nature of thought without attending to such constraints. There are many motivations for such a project, some highly respectable, some less so. Among the respectable motivations we find the belief that the space of possible minds far exceeds the space of biologically possible minds and that investigating the larger, less-constrained space is a key to understanding the special nature of the biological space itself (see, e.g., Sloman 1984). Also among the respectable motivations we find the need to work on isolable, tractable problem domains. It is thus clearly much easier to work on a chess playing algorithm than to try to solve a myriad of evolutionarily basic problems (vision, spatial skills, sensorimotor control, etc.) in an integrated, robust, and flexible fashion. The evolutionary reflections on the nature of human thought can only count against the direction of work in the classical cognitivist tradition if there is some realistic alternative approach. And some cognitive scientists do not see any such alternative (but see part 2 for some reasons for optimism). It is also worth noting, as stressed at the very beginning, that what I am calling the classical cognitivist tradition by no means exhausts the kinds of work already being done even in what I shall later view as conventional artificial intelligence (the intended contrast is with the PDP approach investigated in part 2). Thus, early work on cybernetics, more recent work on low-level visual processing, and some work in robotics can all be seen as attempting to do some justice to the kinds of biological constraints just detailed. To take just one recent example, Mike Brady of Oxford recently gave a talk in which he explained his interest in work on autonomously guided vehicles as rooted in the peculiar task demands faced by such vehicles as robot trucks that must maneuver and survive in a real environment (see Brady et al. 1983). These included severe testing of modules by the real environment, three-dimensional ranging and sensing, real-time sensory processing, data fusion and integration from various sensors, and dealing with uncertain information. Working on autonomously guided vehicles is clearly tantamount to working on a kind of holistic animal micro world: such work is forced to respect many (but not all) of the constraints that we saw would apply to evolved biological systems.

Classical cognitivism tries to make a virtue out of ignoring such constraints. It concentrates on properties *internal* to the individual thinker, paying at best lip service to the idea of processing that exploits the world; it seeks neat, design-oriented, mathematically well understood solutions to its chosen problems; and it chooses its problems by fixating on various interesting high-level human achievements like conscious planning, story understanding, language parsing, game playing, and so forth. Call this a MIND methodology. MIND is a slightly forced acronym meaning: focused on Mature (i.e., evolutionarily recent) achievements; seeking Internalist solutions to information processing problems (i.e. not exploiting the world); aimed at Neat (elegant, well-understood) solutions; and studying systems from an ahistorical, Design-oriented perspective. The methodology of MIND thus involves looking at present human achievements, fixating on various intuitively striking aspects of those achievements (e.g., planning, grammatical competence, creativity), then treating each such high-level aspect as a separate domain of study in which to seek neat, internalist, design-oriented solutions, and hoping eventually to integrate the results into a useful understanding of human thought. This general strategy is reflected in the plan of AI textbooks, which will typically feature individual chapters on, e.g., vision, parsing, search, logic, memory, uncertainty, planning, and learning (this is the layout of Charniak and McDermott 1985). Our earlier reflections (sections 2 to 4) already give us cause to doubt the long-term effectiveness of such a methodology, at least if the goal is an understanding of *human* thought.

Such reflections suggest two particular, related pitfalls for MIND-style theorizing. At the risk of repetition, I should like to end this chapter by making them as explicit as possible. The first is what I shall call the danger of missing the mess. Human thought in evolutionarily recent tasks (like chess playing and probabilistic reasoning) may be very messy indeed. We may be placed very much in the role of a tinkerer, having to make do with processing powers and strategies designed for much more basic tasks. A pure, design-oriented approach to these recent achievements seems to be, we saw, in severe danger of missing the mess. Worse still, it may be that what makes us messy probabilistic reasoners, say, (compulsive pattern completing, holistic information processing, etc.) is the very thing that makes us flexible and creative in our capacity to *use* such reasoning in the service of our changing needs and desires (on our probabilistic-reasoning skills see, e.g., Kahneman et al. 1982). In such cases, missing the mess would not merely yield a psychologically incorrect account of our thought, it would blind us to the correct explanation of further aspects of our thought. The point about mess holds at much-lower levels too. With unusual clarity one writer recently wrote: "A computer scientist contemplating a monkey's central nervous system (or that of a human—the differences are slight) could be forgiven for wondering whether it was designed by a genius or a lunatic.... Only a genius could design something so effective. Only a maniac could design something so complicated. These contradictory features could only co-exist in something that was not designed at all, but simply evolved" (Durham 1987, 28).

The second pitfall (call it "bad focus") is in some ways just a different kind of perspective on the first. It again points to the danger of a methodology that fixates on intuitively striking, recent cognitive achievements. But the worry this time is that some such achievements may not be proper objects of computational investigation at all. It is not easy to convey the flavor of this possibility in advance of some of the detailed work in part 2 but an analogy drawn from evolutionary biology may help.

An adaptationist in evolutionary biology is one who holds that natural selection is an optimizing process (subject to trade-offs), that the biological organism consists of a set of traits (aggressiveness, brain size, leg length, etc.) that are individually optimized (subject to trade-offs), that the presence of any given trait is to be explained by alluding to the biological advantages accruing to those who exhibit it (adapted from Gould and Lewontin 1978). In short, an adaptationist is someone who seeks a direct evolutionary explanation for *every* striking feature of an organism, i.e., they seek to explain the presence of the feature by telling a story about the selective advantages it confers. But not every striking feature *has* any such direct function. The shape of the chin, to use a classic example, is merely a striking by-product, the result of an architectural relation that obtains between anatomical features selected on quite independent grounds.

In a seminal paper Gould and Lewontin (1978) caricature the adaptationist approach by applying such reasoning to two nonbiological cases. The first concerns the spandrels of Saint Marks Cathedral in Venice. A spandrel is a triangular space formed by the intersection of two rounded arches. Spandrels are a necessary structural by-product caused by mounting a dome on a number of rounded arches. The spandrels of Saint Marks, however, have been put to particularly good use, as can be seen in figure 4.3. The spandrels are used to express the Christian themes of the dome. In this case a man (said to represent one of the biblical rivers) is seen pouring water from a pitcher. Overall, the effect of the designs worked into the spandrels is so striking that we might even tempted to view the overall structure of pillars and dome as themselves a result of the need to have a triangular space for the designs. But this, of course, would be precisely to reverse the true order of explanation. As a *result* of the decision to rest a dome on rounded arches, spandrels come into being as an inevitable by-product. These were then exploited by the artist or designer.

As a second example consider the ceiling of Kings College Chapel (Gould and Lewontin 1978, 254). The ceiling (described by Wordsworth as "that branching roof self-poised, and scooped into ten thousand cells, where light and shade repose") is supported by a series of pillars that are fan-vaulted at the top. Where the fan vaultings meet between pillars, a star-shaped space is inevitably created. In Kings College Chapel these star-shaped spaces have been decorated with portcullises and roses (see figure 4.4). Once again, the architectural constraint is clearly the main source of the design. Gould and Lewontin point out, "Anyone who tried to argue that the structure exists because the alternation of rose and portcullis makes so much sense in a Tudor chapel would be inviting ... ridicule" (1978, 254).

In the cases cited, we would regard the adaptationist explanations of spandrels and star-shaped spaces as bizarre. The lesson, then, is that we must not simply accept our intuitions about the basic and central features of an organism as a reliable guide to its decomposition into a set of individual traits in need of adaptationist explanation. To do so is to commit a fallacy of reification. In this fallacy intuitively striking but emergent features of a complex object are reified, and direct explanations of each feature are constructed. By trying to deal with high-level cognitive features without first attending to their basic underpinnings, standard AI, I believe, commits a version of this fallacy. Table 4.1 makes the parallel explicit. The fact that the features listed to the right of human thought strike us as suitable individual objects of computational investigation may be an effect

Figure 4.3
One of the spandrels of Saint Mark's. Reproduced by permission from Gould and Lewontin 1978, 582

Figure 4.4
The ceiling of King's College chapel. Reproduced by permission from Gould and Lewontin 1978, 583

Table 4.1
The parallel between architecture and cognition

Complex object	Striking features
Saint Marks	spandrels
Kings College chapel	star shapes
human thought	learning
	induction
	word recognition
	reaching
	frame-based reasoning
	memory
	planning
	rule following

of our linguistic perspective, much as the isolation of the spandrels is an effect of an artistic perspective. Having a language in which sentential formulations pick out various kinds of mental states may mislead us into believing that these sentential formulations pick out computationally isolable achievements suitable for MIND-style explanation. But such explanation may project properties of the sentential formulations back into the heads of those who use sentential formulations. As we saw in chapter 3, sentential formulations are used to do many things, none of which need involve tracking computationally or neurophysiologically isolable states of the head. Such back projection is as unwarranted as it is evolutionarily implausible. (For a thoroughgoing attack on the sentential approach see Churchland 1986.) Being may surely come to use language without what goes on in their heads having the properties of the language they come to use. Because we are so steeped in language, its apparatus may mislead us to conceive of the computational and neural substrate of thought in terms of the categories that language uses to describe thought. But such a policy (pursued, we saw, by standard functionalists and classical cognitivists alike) has begun increasingly to look ill advised.

In the remaining half of this book I look at a different approach, variously known as connectionism and parallel distributed processing. This approach, it seems to me, goes some way toward avoiding the gross biological and philosophical implausibility of much conventional work in AI and cognitive science.

II

The Brain's-Eye View

The rich behaviour displayed by cognitive systems has the paradoxical character of appearing on the one hand tightly governed by complex systems of hard rules, and on the other to be awash with variance, deviation, exception, and a degree of flexibility and fluidity that has quite eluded our attempts at simulation.... The subsymbolic paradigm suggests a solution to this paradox.

—Paul Smolensky, "On the Proper Treatment of Connectionism"

Chapter 5
Parallel Distributed Processing

1 PDP or Not PDP?

Classical cognitivism was indeed the pure science of the mind. Or perhaps it was the pure science of the mind's own idea of the mind. However you see it, it would be hard to exaggerate its deliberate intellectual distance from the messy substrata of biological fact. On the structure of the brain and the phylogeny of cognitive processes, the classical cognitivist maintained a studied indifference. But perhaps cognitive science cannot afford that indifference.

Parallel distributed processing (or connectionism) is an attempt to provide slightly more biologically realistic models of mind. Such models, though hardly accurate biologically, are at least *inspired* by the structure of the brain. Moreover, they are tailored, in a sense to be explained, to evolutionarily basic problem-solving needs, like perceptual pattern completion. These models, I shall argue, offer the best current prospect for soothing the philosophical and biological sore spots inflamed (I hope) by the first half of this book.

As counterpoint to the enthusiasm, a word of warning. There is a certain danger in the extreme polarization of cognitive science which this treatment may seem to imply. The danger is summed up in the slogan "PDP or not PDP?" This, I hope to show, is *not* the question. PDP is not a magic wand.[1] And a connectionist touch will not turn a Von Neumann frog into a parallel distributed princess. Both our deeper explanatory understanding of cognition and on some occasions our actual processing strategies may well *demand* the use of higher level, symbolic, neoclassical descriptions. For all that, connectionist models offer insights into the way nature may provide for certain properties that seem to be quite essential to what we consider intelligent thought. Walking this tightrope between the cognitivist and connectionist camps is a major task of the next five chapters. First though, we had better get some idea of what a PDP approach actually involves.

2 *The Space between the Notes*

The musician's talent, it is sometimes said, lies not in playing the notes but in spacing them. It is the silences that makes the great musician great. As it is with music, so it is with connectionism. The power of a connectionist system lies not in the individual units (which are very simple processors) but in the subtly crafted connections between them. In this sense such models may be said to be examples of a brain's eye view. For it has long been known that the brain is composed of many units (neurons) linked in parallel by a vast and intricate mass of junctions (synapses). Somehow this mixture of relatively simple units and complex interconnections results in the most powerful computing machines now known, biological organisms. Work in parallel distributed processing may be said to be neurally inspired in the limited sense that it too deploys simple processors linked in parallel in intricate ways. Beyond that the differences are significant. Neurons and synapses are of many different types, with properties and complexities of interconnection so far untouched in connectionist work. The PDP "neuron" is a vast simplification. Indeed, it is often unclear whether a single PDP unit corresponds in any useful way to a single neuron. It may often correspond to the summed activity of a group of neurons. Despite all the differences, however, it remains true that connectionist work is closer to neurophysiological structure than are other styles of computational modeling (see Durham 1987; McClelland, Rumelhart, and the PDP Research Group 1986, vol. 2, chapters 20–23).

Neurally inspired theorizing has an interesting past. In one sense it is a descendant of gestalt theory in psychology (see Köhler 1929; Baddeley 1987). In another more obvious sense it follows the path of cybernetics, the study of self-regulating systems. Within cybernetics the most obvious antecedents of connectionist work are McCulloch and Pitts 1943; Hebb 1949; and Rosenblatt 1962. McCulloch and Pitts demonstrated that an idealized net of neuronlike elements with excitory and inhibitory linkages could compute the logical functions *and*, *or*, and *not*. Standard results in logic show that this is sufficient to model any logical expression. Hebb went on to suggest that simple connectionist networks can act as a pattern-associating memory and that such networks can teach themselves how to weight the linkages between units so as to take an input pattern and give a desired output pattern. Roughly, Hebb's learning rule was just that if two units are simultaneously excited, increase the strength of the connection between them (see McClelland, Rumelhart, and the PDP Research Group 1986, vol. 1, p. 36, for a brief discussion). This simple rule (combined with an obvious inhibitory variant) is not, however, as powerful as those used by modern day connectionists. Moreover, Hebb's rules were not sufficiently rigorously expressed to use in working models.

This deficiency was remedied by Rosenblatt's work on the so-called perceptron. A perceptron is a small network of input units connected via some mediating units to an output unit. Rosenblatt's work was especially important in three ways, two of them good, one disastrous. The two good things were the use of precise, formal mathematical analysis of the power of the networks and the use of digital-computer simulations of such networks (see McClelland, Rumelhart, and the PDP Research Group 1986, vol. 1, p. 154–156). The disastrous thing was that some overambitious and politically ill advised rhetoric polarized the AI community. The rhetoric elevated the humble perceptron to the sole and sufficient means of creating real thought in a computer. Only by simulating perceptrons, Rosenblatt thought, could a machine model the depth and originality of human thought.

This claim and the general evangelism of Rosenblatt's approach prompted a backlash from Minsky and Papert. Their work *Perceptrons* (1969) was received by the alienated AI community as a decisive debunking of the usurping perceptrons. With the rigorous mathematical analysis of linear threshold functions Minsky and Papert showed that the combinatorial explosion of the amount of time needed to learn to solve certain problems undermined the practical capacity of perceptronlike networks to undergo such learning. And they further showed that for some problems no simple perceptron approach could generate a solution. Rather than taking these results as simply showing the limits of one type of connectionist approach, Minsky and Papert's work (which was as rhetorically excessive as Rosenblatt's own) was seen as effectively burying connectionism. It would be some years before its public resurrection.

But the miracle happened. A recent three-page advertisement in a leading science journal extols the slug as savant, claiming that the parallel neural networks of the slug suggest powerful new kinds of computer design. The designs the advertisers have in mind are quite clearly based on the work of a recent wave of connectionists who found ways to overcome many of the problems and limitations of the linear-thresholded architectures of perceptrons. Landmarks in the rise of connectionism include Hinton and Anderson 1981; McClelland and Rumelhart 1985a; and McClelland, Rumelhart, and the PDP Research Group 1986. Other big names in the field include J. Feldman, D. Ballard, P. Smolensky, T. Sejnowski, and D. Zipser. It would be foolhardy to attempt a thorough survey of this extensive and growing literature here. Instead, I shall try to convey the flavor of the approaches by focusing on a few examples. These have been chosen to display as simply as possible some basic strategies and properties common to a large class of connectionist models. The precise algorithmic form of such models varies extensively. The *emergent* properties associated with the general class of models bear the philosophical and biological weight. This is reflected in the discussion that follows.

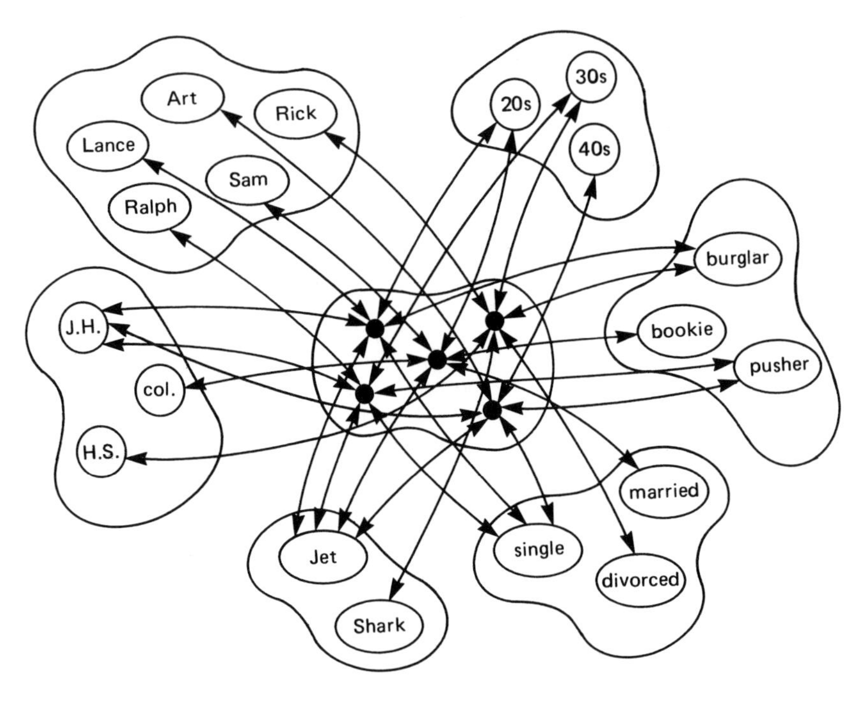

Figure 5.1
A local hardwired network. From McClelland, Rumelhart, and Hinton 1986, 28

3 *The Jets and the Sharks*

Following an example developed in McClelland, Rumelhart, and Hinton 1986,[2] let us imagine two New York street gangs, the Jets and the Sharks. Some of the facts about them are presented in table 5.1. One way to encode and store this kind of information is with a local hard-wired network, which may be represented as in figure 5.1. The following conventions are adopted:

- Irregular clouds signify the existence of mutual inhibitory links between all the units within a cloud. Thus, figure 5.2 is composed of three units, one signifying that the individual concerned is in his twenties, another signifying that an individual is in his thirties, and so on. Since no one can be both in his twenties and in his thirties or forties, the units are set up to be mutually inhibitory. If one fires it will dampen the other two.
- Lines with arrow heads represent excitatory links. If the line has an arrow at each end, the link is mutually excitatory. Thus, suppose all burglars are in their thirties. There would be an excitatory link be-

Table 5.1
The Jets and the Sharks

Name	Gang	Age	Education	Marital status	Occupation
Art	Jets	40s	J.H.	sing.	pusher
Al	Jets	30s	J.H.	mar.	burglar
Sam	Jets	20s	col.	sing.	bookie
Clyde	Jets	40s	J.H.	sing.	bookie
Mike	Jets	30s	J.H.	sing.	bookie
Jim	Jets	20s	J.H.	div.	burglar
Greg	Jets	20s	H.S.	mar.	pusher
John	Jets	20s	J.H.	mar.	burglar
Doug	Jets	30s	H.S.	sing.	bookie
Lance	Jets	20s	J.H.	mar.	burglar
George	Jets	20s	J.H.	div.	burglar
Pete	Jets	20s	H.S.	sing.	bookie
Fred	Jets	20s	H.S.	sing.	pusher
Gene	Jets	20s	col.	sing.	pusher
Ralph	Jets	30s	J.H.	sing.	pusher
Phil	Sharks	30s	col.	mar.	pusher
Ike	Sharks	30s	J.H.	sing.	bookie
Nick	Sharks	30s	H.S.	sing.	pusher
Don	Sharks	30s	col.	mar.	burglar
Ned	Sharks	30s	col.	mar.	bookie
Karl	Sharks	40s	H.S.	mar.	bookie
Ken	Sharks	20s	H.S.	sing.	burglar
Earl	Sharks	40s	H.S.	mar.	burglar
Rick	Sharks	30s	H.S.	div.	burglar
Ol	Sharks	30s	col.	mar.	pusher
Neal	Sharks	30s	H.S.	sing.	bookie
Dave	Sharks	30s	H.S.	div.	pusher

Source: McClelland, Rumelhart, and Hinton 1986, p. 27, fig. 10.

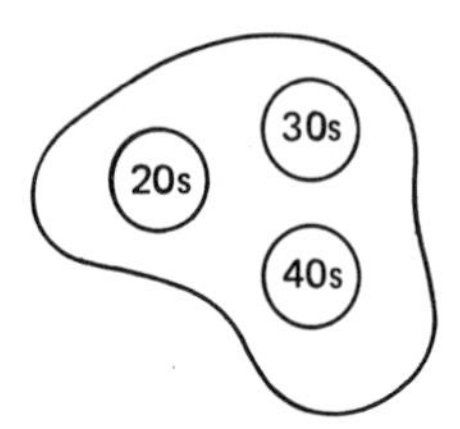

Figure 5.2
Inhibitory links

tween each burglar unit and the thirties unit. If, in addition, *only* burglars are in their thirties, the thirties unit would be excitatorily linked to the units representing burglars.

- Solid black spots signify individuals and are connected by excitatory links to the properties that the individual has, e.g., one such unit is linked to units representing Lance, twenties, burglar, single, Jet, and junior-high-school education.

By storing the data in this way, the system is able to buy, at very little computational cost, the following useful properties: content addressable memory, graceful degradation, default assignment, and generalization. I shall discuss each of these in turn.

Content addressable memory

Consider the information that the network encodes about Rick. Rick is a divorced, high-school educated burglar in his thirties. In a more conventional approach this information would be stored at one or several addresses, with retrieval dependent upon knowing the address. But a designer may want to make all this information accessible by *any* reasonable route. For example, you may know only that you want data on a Shark in his thirties, or you may have a description that is adequate to identify a unique individual but nevertheless contains some errors. Such flexible (and in this case error-tolerant) access to stored information is known as content addressable memory. Humans certainly have it. To borrow McClelland and Rumelhart's lovely example, we can easily find the item that satisfies the description: "is an actor, is intelligent, is a politician," despite the meagre and perhaps partially false description. Flexible, error-tolerant access requires some computational acrobatics in a conventional system. In the *absence* of errors, a technique called hash coding is quite efficient (see Knuth 1973). The error-tolerant case, however, requires an expensive best-match search. Storing the information in a network of the kind just described is a very natural, fast, and relatively cheap way of achieving the same result.

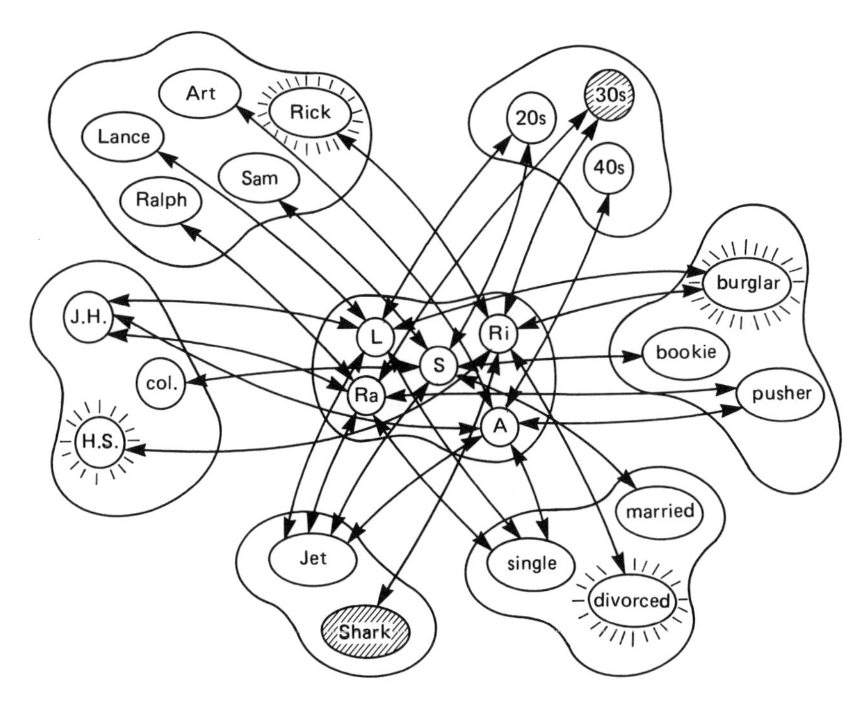

Figure 5.3
The pattern of activation for a Shark in his thirties. Hatching = input activations. Sunburst = units to which activation spreads. The diagram is based on McClelland, Rumelhart, and Hinton 1986, p. 28, fig. 11.

It is easy to see how this works. Suppose you want to know who satisfies the description "is a Shark in his thirties." The thirties and Shark units are activated and pass positive values to the units to which they have excitatory links. There is a chain of spreading activation in which first the individual-signifying unit, then those other units to which it is excitatorily linked get activated. The result is a pattern of activation involving the units for Shark, thirties, burglar, divorced, a high-school education, and Rick. The process is shown in figure 5.3. The important point is just this: the same final pattern of activation (i.e., the overall pattern of units active after the spread of activation) could have been achieved by giving the system any one of a number of partial descriptions, e.g., the inputs "Shark, high-school education," "Rick, thirties," and so on. Simply by using a network representation of the data, we obtain a flexible, content-addressable memory store.

Graceful degradation

Graceful degradation, as we saw in chapter 4, comes in two, related varieties. The first demands that a system be capable of sustaining some hardware damage without being totally incapacitated. The second demands that a system be capable of behaving sensibly on the basis of data that is partial or includes errors. The PDP mode of data storage and retrieval supports both these qualities. The capacity to tolerate some hardware damage is best seen in more distributed networks of the kind detailed in subsequent examples. Sensible behavior despite data that is partial or includes errors can be demonstrated in a local net. We have already seen how partial data can prompt a full pattern of activation. The extension to data that includes errors is easy.

Suppose we want to retrieve the name of an individual whom we believe to be a Jet, a bookie, married, and educated at the junior-high-school level. In fact, no one in our model satisfies that description. The best fit is Sam, who is a bookie, a Jet, and married but has a college education. The network can cope, thanks to the *inhibitory* links. The system works like this. The units for bookie, married, Jet, and (the mistake) a junior-high-school education are activated. The units for bookie and married directly excite only one of the units that specify individuals. (I have labeled it in figure 5.4.) The Jet unit excites the individual-signifying units labeled *A, S, Ra,* and *L*. (Only Rick, whose individual-signifying unit is labeled *ri*, is a Shark.) The junior-high-school-education unit excites *L, Ra* and *A*. In sum:

The bookie unit excites *S*.

The married unit excites *S*.

The Jet unit excites *A, S, Ra, L*.

The J.H.-education unit excites *L, Ra, A*.

Thus, the *S* unit is stimulated by three times, and the *L*, and *Ra*, and *A* units twice. But the various individual-representing units are *themselves* connected in a mutually inhibitory fashion, so the strong, threefold activation of the *S* unit will tend to inhibit the weaker, twofold activation of the *A, L,* and *Ra* units. And when the activation spreads outwards from the individual units, the *S* unit will pass on the most significant excitatory value. The *S* unit is excitatorily linked to the name unit Sam. And the various name units too, are competitively connected via mutually inhibitory links. Thus "Sam" will turn out to be the network's chosen completion of the error-involving description beginning "Jet," "bookie," "married," "junior-high-school education." A sensible choice. The spread of activation responsible is shown in figure 5.4.

Default assignment

Suppose that you don't know that Lance was a burglar. But you *do* know that most of the junior-high-school-educated Jets in their twenties are burglars rather than bookies or pushers (see the data table in figure 5.2). It

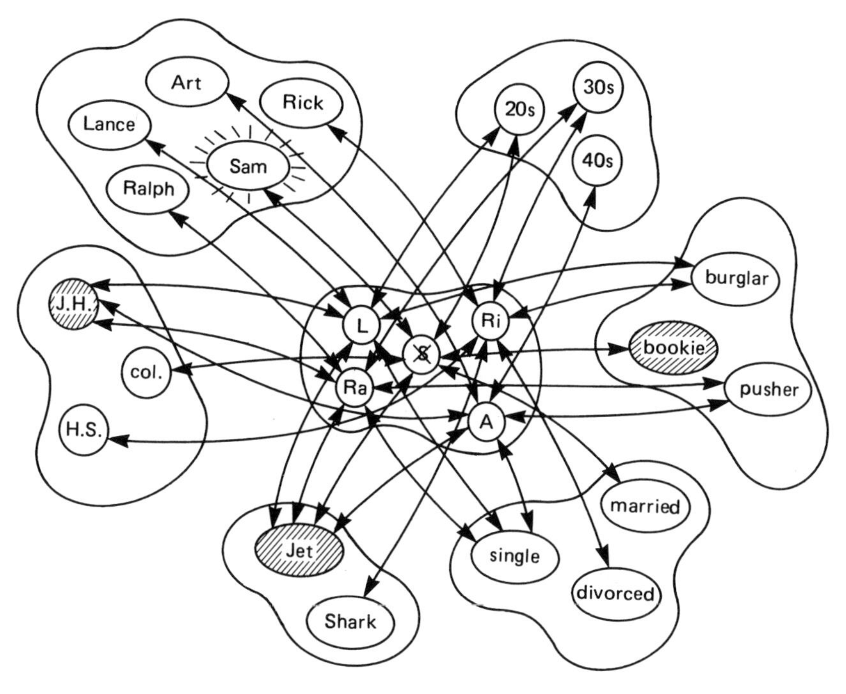

Figure 5.4
The pattern of activation for a Jet bookie with a junior-high-school education. A unit labeled with a gang member's initial stands for that individual. The input patterns are marked with hatching. The strongly activated individual unit is marked with an x, and the name unit it excites is marked with a sunburst. The diagram is based on McClelland, Rumelhart, and Hinton 1986, p. 28, fig. 11.

is reasonable to assume that Lance too is a burglar, at least until we learn otherwise. This kind of assumption is called a default assignment. It is generally good practice to assume that patterns found in known data will extend to cover new cases. The network under consideration is able to assign default values in this way. Unfortunately, it would be too messy to attempt a diagrammatic representation of this process. But the basic story is not hard to grasp. Suppose that we do not know that Lance is a burglar. Still, when we activate the name unit Lance it will activate the units for all of Lance's known properties (i.e., Jet, junior-high-school education, married, in his twenties). These property units in turn will excite the units for others who have these properties. If most of those who share Lance's *known* properties also share a particular further property (i.e., if there is a real pattern here) then the spread of activation from these units will combine to activate for Lance the unit representing the further property in question. In this way the burglar unit gets activated as a kind of default assignment to Lance.

Flexible generalization
The property of flexible generalization is closely related to those considered above. Indeed, in a number of respects we may look upon *all* the properties treated in this example as involving different high-level descriptions and uses of the same underlying computational strategy of pattern completion. In this final case, the pattern-completing talent of the system is used to generate a *typical* set of properties associated with some description, even though all the system directly knows about are individuals, none of whom need be a perfectly typical instantiation of the description in question. Thus, suppose we seek a sketch of the typical Jet. It turns out, as we saw in the discussion of default assignments, that there are indeed patterns in Jet membership, although no individual Jet is a perfect example of all these patterns at once. Thus the majority of Jets are single, in their twenties and educated to the junior-high-school level. No significant patterns exist to specify particular completions of a pattern beginning "Jet" along the other dimensions that the system knows about. Thus, if the system is given "Jet" as input, the units for single, twenties, and junior-high-school education will show significant activity, while the rest will mutually cancel out. In this way, the system effectively generalizes to the nature of a typical Jet, although no individual Jet in fact possesses all three properties simultaneously.

Perhaps what is most significant here is not the capacity to generalize per se so much as the *flexibility* of the generalizing capacity itself. As McClelland and Rumelhart point out, a conventional system might explicitly create and store various generalizations. One striking feature of the PDP version is its capacity to generalize in a very flexible way with no need for any explicit storage or prior decisions concerning the form of required generalizations. The network can give you a typical completion of any pattern you care to name if there *is* some pattern in the data. Thus, instead of asking for details of a typical Jet, we could have asked for details of a typical person in his twenties educated at the junior-high-school level or a typical married pusher, and so on. The network's generalization capacity is thus flexible enough to deploy available data in novel and unpredicted ways, ways we needn't have thought of in advance. As our account progresses, this kind of unforced flexibility will be seen to constitute a major advantage of PDP knowledge representation.

4 Emergent Schemata

In chapter 2 (especially sections 2 and 3) we spoke of scripts and schemata. These are special data structures that encode the stereotypic items or events associated with some description. For example, Schank and Abelson's restaurant script held data on the course of events involved in a typical visit to a restaurant. The motivation for introducing scripts and schemata was simple. Human reasoning involves filling in numerous default values in order to understand what we see or are told. The schema is meant to capture this background knowledge and (by associating it with a particular description, e.g., "restaurant visit," "birthday party," etc.) enable us to deploy the right subset of our total background knowledge at the right time. But as we saw, the traditional schema-based approach was beset by difficulties. Individual scripts or schemata proved too structured and inflexible to be able to cope with all the variants of a situation (e.g., novel situations, mixtures of stereotypic scenarios, etc.), yet humans are sensible enough to take such variants in stride. The obvious way of overcoming this problem is to accept a massive proliferation of schemata. But the computational expense would be prohibitive. As Dreyfus pointed out, there is no obvious end to the multiplication of explicit schemata required.

McClelland, Rumelhart, and the PDP Research Group (1986, vol. 2, p. 20–38) detail a PDP model in which the properties of explicit stored schemata emerge simply from the activity of a network of units that respond to the presence or absence of microfeatures of the schemata in question. These emergent schemata are presented as a partial solution to the dilemma facing classical approaches that depend on explicit stored schemata, the dilemma that schemata are the structure of the mind on the one hand and schemata must be sufficiently malleable to fit around most everything on the other (McClelland, Rumelhart, and the PDP Research Group 1986, vol. 2, p. 20). At an abstract level (we shall get more concrete in a moment) McClelland, Rumelhart et al. characterize the kind of malleability required as involving the capacity of a system to adjust the default values of each item in a scenario in a way that is sensitive to all the data known in a current situation. Thus, it might be reasonable to assign "some meat" as a default value for the slot "contents of refrigerator." But if you are also told that the refrigerator belongs to a vegetarian, this default assignment should change. This kind of flexibility has to be explicitly provided for in any conventional schema-based approach, at a cost of significant information-processing complexity.

The PDP model gets around these difficulties in a natural and unforced way by not having explicit schemata represented at all. As the PDP Research Group put it: "Schemata are not 'things' (in this model). There is no representational object which is a schema. Rather, schemata emerge at the moment they are needed from the interaction of large numbers of much simpler elements all working in concert with one another" (McClelland, Rumelhart, and the PDP Research Group 1986, vol. 2, p. 20). The key to the system, as they note, is the notion of subpatterns of units that tend to come on together, because of excitatory links. The tendency can be overwhelmed by sufficient incompatible excitation and inhibition from elsewhere. But insofar as it exists, it acts like a stored schema. Yet there is no need to decide in advance on a set of schemata to store. Instead, the system learns (or is told) patterns of cooccurrence of individual items making up the schema, and the rest (as we shall now see) comes in the back door when required.

The particular model detailed by McClelland, Rumelhart, et al. concerns our understanding of a typical room, e.g., a typical kitchen or a typical office.[3] The general interest of the model for us is twofold. First, it moves us in the direction of more *distributed* representations. Second, it shows how high-level symbolic characterizations (e.g., our idea of a typical kitchen) can be among the emergent properties of a network of simpler entities and how such high-level descriptions ("believes that ...") may be perfectly *correct* without indicating the underlying computational structure of the system so described. This recalls our account of the emergence of spandrels from the interaction of other, more basic, architectural features, and it will prove important later on.

Room-dwelling human beings have ideas about the likely contents of particular types of rooms. If you are told to imagine someone in the kitchen standing by the cooker, you may well fill in some other details of the room. If you did, there would likely be a refrigerator and a sink, some wall cupboards, and so on. How does this happen? One answer would be to propose a "mental file" marked "kitchen" and detailing all the expected items. But that approach brings in its wake all the difficulties raised earlier. The answer we are being asked to consider goes like this. You were exposed to the contents of lots of rooms. You saw objects clustering on these occasions. Generally, when you were in a room with a cooker, you were in a room with a sink and not in a room with a bed. So suppose you had something like a set of PDP units responding to the presence of particular household items (this is clearly an oversimplification, but it will do for current purposes). And you fixed it so that units that tended to be on together got linked by an excitatory connection and units that tended to go on and off independently got linked by an inhibitory connection. You would get to a point where activating the unit for one prominent kitchen feature (e.g., the cooker) would activate in a kind of chain all and only the units standing for items commonly found in kitchens. Here, then, we have an emergent schema in its grossest form. Turn on the oven unit and after a while you get (in the particular simulation done by McClelland, Rumelhart et al.) oven, ceiling, walls, window, telephone, clock, coffee cup, drapes, stove, sink, refrigerator, toaster, cupboard, and coffeepot. So far, all is conventional. But let's look more closely at some of the properties of this mode of representing the data. The first interesting property is the distributed nature of the system's representation of a kitchen. The concept of a kitchen here involves a pattern of activation over many units that may stand for (or respond to) more basic items treated in our daily language, or even for items not visible to daily talk at all. Such features might be functional or geometrical properties of objects. Whether or not the low-level features (known as microfeatures) are visible to daily talk, the strategy of building functional correlates of high-level concepts out of such minute parts in a PDP framework brings definite advantages. The main one, which I shall reserve for treatment in the next chapter, is the capacity to represent fine shades of meaning. A spin-off is the capacity of the gross, high-level concept or ability to degrade gracefully in the first way mentioned in the previous example. That is, it turns out to be robust enough to withstand some damage to the system in which it is distributively encoded. Thus, suppose the coffeepot unit or its links to other units got destroyed. The system would still have a functional kitchen schema, albeit one lacking a coffeepot default.

The distinction between a local and a distributed representational system is somewhat perspectival.[4] Distribution is in the eye of the beholder or at best in the functional requirements of the system itself. Thus, in the example described, we have a distributed (hence, gracefully degradeable) representation of kitchens but a local and quite gracelessly degradable representation of coffeepots. Of course, we could in turn have a distributed representation of coffeepots (e.g., as a pattern of activation across a set of units standing for physical and functional features of coffeepots). But there will always be *some* local representation at the bottom line; even if it lies well below the level of any features we might fix on in daily talk. Conversely, even an intuitively local network like the Jets and the Sharks net described in section 3 can be seen as a distributed representation of some slightly artificial concept such as Jet-Shark membership. The second and, I think, more interesting property of the room network is the multiplicity and flexibility of the emergent schemata it supports. McClelland, Rumelhart, et al. ran a simulation involving forty names of household features. After fixing connectivity strengths according to a rough survey of human opinions, they found that the network stored five basic schemata, one for each type of room they had elicited opinions on. (They were: kitchen, office, bedroom, bathroom, and living room.) The sense in which it stored five basic patterns was that if you set just *one* description unit on (e.g., the oven or bed unit) the system would always settle on one of five patterns of activation. But the system proved *much* more flexible than this state of affairs initially suggests. For many other final patterns of activation proved possible if more than one description was turned on. In fact, there were 2^{40} possible states into which it might settle, one for each vertex of a 40-dimensional hypercube. It is just that some of these points are easier to reach than others. It will thus find a "sensible" pattern completion (subject to how sensible its stored knowledge is) even for the input pattern "bed, bath, refrigerator." To show this in action, McClelland, Rumelhart, et al. describe a case where "bed, sofa" is given as the input pattern. The system then sets about trying to find a pattern of activation for the rest of its units that respects, as far as possible, all the various constraints associated with the presence of a bed and a sofa simultaneously. What we want is a set of values that allows the final active schema to be sensitive to the effects of the presence of a sofa on our idea of typical bedroom. In effect, we are asking for an unanticipated schema: a typical bedroom-with-a-sofa-in-it. The system's attempt is shown in figure 5.5. It ends up having chosen "large" as the size description (whereas in its basic bedroom scenario the size is set at "medium") and having added "easy-chair." "floor lamp," and "fireplace" to the pattern. This seems like a sound choice. It has added a subpattern strongly associated with sofa to its bedroom schema and has adjusted the size of the room accordingly.

Here, then, we have a concrete example of the kind of flexible and sensible deployment of cognitive resources that characterizes natural intelligence. Emergent schemata obviate the need to decide *in advance* what possible situations the system will need to cope with, and they allow it to adjust its default values in a way maximally consistent with the range of possible inputs. Just this kind of natural flexibility and 'informational holism' constitutes, I believe, a main qualitative advantage of PDP approaches over their conventional cousins.

5 *Distributed Memory*

My final example concerns the process of learning and remembering. It follows work originally presented in McClelland and Rumelhart 1985 and modified in volume 2 of McClelland, Rumelhart, et al. The goal of the work is to generate a model of memory in which the storage of traces of specific experiences gives rise in a very natural way to a general, nonspecific understanding of the nature of the domain in question. Thus, for example, storage of traces of specific experiences of seeing dogs will give rise to a general, prototypical idea of doghood. This neatly sidesteps a recurrent problem in modeling memory, namely, the choice between representing specific and general information. In terms of its behavior, the model looks as if it explicitly generates and stores prototypes (of e.g., the typical dog). But as in the schema example, there are no such explicit, stored items. Instead, the prototype-based understanding is an emergent property of the system's way of dealing with specific experiences. The model shares many of the features of our two previous examples, but it enables us to extend our discussion to include:

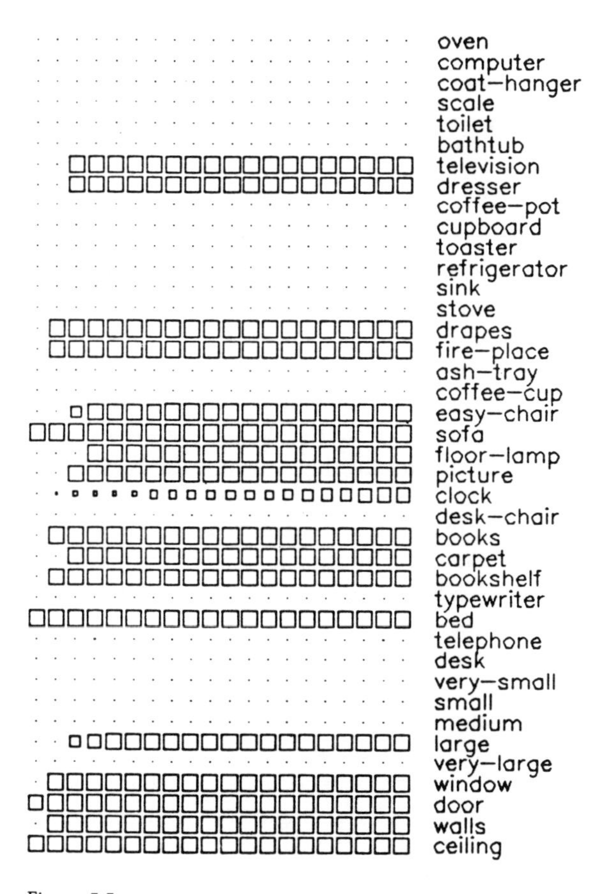

Figure 5.5
The output of the room network with *bed, sofa,* and *ceiling* initially clamped. The result may be described as a large, fancy bedroom. White boxes indicate active units. The vertical sets of such boxes, reading from left to right, indicate the successive states of activation of the network. The starting state (far left) has the bed, sofa, and ceiling units active. The figure is from Rumelhart, Smolensky, et al., 1986, 34.

- the use of learning rules in PDP,
- the economy of PDP storage (so-called "superpositional storage"),
- the PDP capacity to mimic the explicit storage of rules, prototypes, and so on,
- the relation of PDP to experimental psychological data, and
- the limits of networks without hidden units.

The model that McClelland and Rumelhart propose is a fairly standard PDP network of the kind discussed above. The network is exposed to successive sets of inputs given in terms of a fixed set of representational primitives, i.e., a fixed set of features to which (on this interpretation) its units (or sets of units) are seen as sensitive. Such features may include visual ones, like color or size, and nonvisual ones, like names. As a model of memory the task of the system is just this: given some input with features $f_1, \ldots, f_n$ (say, $f_1, \ldots, f_{10}$ for definiteness), the system needs to store the input in a way that enables it later to re-create the input pattern from a fragment of it acting as a cue. Thus, if the system is given values for $f_1, \ldots, f_4$, we want it to fill in $f_5, \ldots, f_{10}$ with values somehow appropriate to its earlier experience in which $f_1, \ldots, f_{10}$ were active. A simple learning rule, called the delta rule, suffices to produce this kind of behavior. The delta rule is explained formally in McClelland, Rumelhart, et al. 1986, vol. 1, chapters 2, 8, 11, and 2, chapter 17. Informally, it works like this. Getting a system to re-create an earlier activation pattern $f_1, \ldots, f_{10}$ when given the fragment $f_1, \ldots, f_4$ amounts to requiring that the *internal* connections between the units in the net be fixed so that activation of the fragment $f_1, \ldots, f_4$ causes activation of the rest of the pattern $f_1, \ldots, f_{10}$. So we need strong excitatory links between $f_1, \ldots, f_4$ and the units $f_5, \ldots, f_{10}$. After the system has received a teaching input of $f_1, \ldots, f_{10}$ the delta rule simply gets the system to check whether the internal connections between the units that were active would support such a re-creation. If they would not—as is generally the case—it modifies the overall pattern of connectivity in the required direction. The delta rule is strong enough to guarantee that subject to certain caveats (more on which later) a system following it will learn to re-create the pattern over its units from a suitable fragment of that pattern.

To get a better idea of how this looks in action, let's consider McClelland and Rumelhart's example of learning about dogs. The example is in many ways similar to previous ones, but it will help bring out the main points nevertheless. First, circumscribe the domain by fixing on a prototypical dog. Take a picture of this, and describe it in terms of a fixed set of representational primitives, say, sixteen features. Next, create a set of specific dog descriptions none of which quite matches the prototype. These are obtained by varying any one of the features of the prototypical dog at random. Now give each individual dog a name. For each dog code its name as a pattern of activation across eight units. Give the network a series of experiences of individual dogs by activating the units that correspond to the dog description and the dog names. After each such exposure, allow the system to deploy the delta rule to lay down a memory trace in the form of a pattern of altered connectivity and to facilitate recall of the last dog description.

After 50 such runs the system had never been exposed to the prototypical dog but only to these distorted instances. The system was then given a fragment of the prototype as input and was able to complete the prototypical pattern. No name units became active, as these, being different for each individual dog, tended to cancel out. Indeed, the network had, in effect, extracted the pattern common to all the slightly distorted inputs, producing a general idea of the prototypical member of the set of which the inputs were instances. Plato scholars will envy the system's ability to see the true form of doghood on the basis of these distorted shadows on the wall of the cave.

The system can do more than extract the prototype from the examples. It can also re-create the pattern of activation for a *specific* dog if it has had a number of exposures to the dog in question and it is given as its recall-prompting input a disambiguating cue, something distinctive about *that* dog: the name of the dog or some distinctive physical feature.

Such, in brief, is the model. Clearly, it does quite well in its main goal of exhibiting prototype knowledge in the absence of any explicit prototype generation and storage procedure. The way of encoding and retrieving specific information results in a functional correlate of prototype-based reasoning. I gave details of a similar phenomenon in section 4 on emergent schemata. This is, in fact, a rather general property of PDP approaches. They exhibit behavior that, taken at face value, strongly suggests a reliance on some special mechanism aimed at the generation and storage of explicit hypotheses concerning the central structures of a domain. But in fact, no special mechanism is required and the hypotheses are not explicitly stored, at least not in any normal sense. In a somewhat parallel case, some classical theorizing about human language acquisition suggest that our linguistic competence arises as the end result of a three-stage process.

(1) We are exposed to a number of utterances.

(2) Perhaps using some innate grammar or just a powerful learning strategy, we seek to formulate at an unconscious level the linguistic rules to account for the structure of the utterances.

(3) We store such rules and deploy them to understand new utterances.

In a PDP model the storage and retrieval strategy targeted on the specific utterances will yield, in much the same way as described above, behavior that looks as if it depends on the formulation and deployment of specific linguistic rules (see, e.g., McClelland, Rumelhart, and the PDP Research Group 1986, vol. 2, chap. 18). But there is no special mechanism required to seek these rules and no need to store them explicitly in advance of some occasion of deployment. It is perhaps misleading to say that the network does not in some sense learn and deploy the rules. For it becomes structured in a way that makes it yield outputs that tend to conform to the rule in a nicely flexible manner. Insofar as rules can *ever* be stored inside a head, or a mechanism, this seems to me to amount to a version of such storage. What is interesting, however, is that such rules depend on no special mechanism of rule generation and storage and are represented in a manner that makes them extremely flexible and sensitive to contextual nuances (more on this in chapter 6). In only this (very important) sense, I believe, do "distributed models ... provide alternatives to a variety of models that postulate abstract summary representations such as prototypes, ... semantic memory representations, or even linguistic rules" McClelland, Rumelhart, et al. 1986, vol. 2, p. 267.

It remains only to mention two further features of the current example and to offer some comments on its limitations. The two further features are superpositional storage and a capacity to model fine-grained experimental data. By "superpositional storage" I mean the property that one network of units and connections may be used to store a number of representations, so long as they are sufficiently distinct (the term often used is "orthogonal") to coexist without confusion. Thus one network can be trained to exhibit behavior appropriate to knowledge of a number of distinct prototypes, such as dog, cat, and bagel (McClelland, Rumelhart, and the PDP Research Group 1986, vol. 2, p. 185). This is because the delta rule can find a set of weights that allows it to complete to quite different patterns according to whether it is given a nonambiguous cue for a dog, cat, or bagel. Interestingly, if it is given an input that is indeterminate between say, a cat and a dog, it will complete to a blended overall pattern, as if it had an idea not just of dogs and cats but also of something halfway between dogs and cats (this property will loom large in later discussions). Indeed, it does have such an idea, insofar as its prototypes come into being only in response to particular calls and so function in a maximally flexible way.

The other remaining feature—the capacity to model fine-grained experimental data—constitutes a major attraction of the approach. By fine-grained experimental data I mean data on the way performance in supposedly central cases is affected by other factors, including context and breakdowns. One of the striking things about some PDP models is that they have ended up, often unintentionally, producing fine-grained behavior of the type found in human performance. This natural modeling of fine-grained data may seem to indicate what Lakatos called a "progressive research programme," i.e., one whose models succeed not just in accounting for the data they set out to explain, but also in predicting or explaining other, perhaps previously undiscovered data as a consequence (see Lakatos 1974). There are various examples of this in the PDP literature. In discussing the case of distributed memory McClelland and Rumelhart mention a number of such points. These include amnesic syndromes, interference phenomena, and blending errors.

Some amnesics, it seems, *can* learn things by very dense repetition of the appropriate experiences. And they prove better at learning general ideas than at remembering specific experiences. Both these traits are neatly explicable in terms of the distributed model. Regarding the first, we need only conjecture that the damage involves a reduction in the *extent* to which each individual experience can effect a change in the connectivity weightings of the network. It follows that amnesics so afflicted can learn, but that they need many more exposures to do so. The second trait naturally follows. Since specific experiences make little impact on the system, it will tend to learn only what is common to a vast number of such experiences. Thus, it will learn general tendencies in preference to any specific individual instances.

Interference phenomena are simple. If a network is trying to recall a pattern *very like* another one it has been exposed to, it may suffer from interference or cross talk. This occurs when units primed to figure in the restoration of one pattern are mobilized by accident in response to a cue for a similar pattern. At first cross talk was seen as a problem for PDP approaches that use superpositional storage. However, it turns out that many of the error patterns that PDP systems are prone to as a result of such interference are very similar to the error patterns of human subjects (see, e.g., the word-recognition and skilled-typing errors dealt with in McClelland, Rumelhart, and the PDP Research Group 1986, vol. 2, p. 139, and vol. 1, p. 14.). Such cross talk is familiar enough. If you want to recall a telephone number and you know another number very like it, your chances of making a mistake greatly increase. In classical models in which the number is simply retrieved from an address associated with a name, there is no reason at all for such a pathology. On a PDP model the pathology falls naturally out of the mode of storage and recall.

Finally, there is the strange phenomenon of blending errors. These involve our making a blend of two memories into a composite whole. Thus, some subjects, when shown a film involving a yellow truck and then given a repetition of the film but with a truck painted blue, later "recall" a *green* truck. This blending is explained as just "the beginnings of the formation of a summary representation," i.e., the start of the process of fixing on the common tendency in a set of input experiences (McClelland, Rumelhart, and the PDP Research Group 1986, vol. 2, p. 208). This, incidentally, gives the lie to the claim that PDP models *don't use* summary representations at all (recall the passage from vol. 2, p. 207 quoted above). McClelland, Rumelhart, et al. cannot have it both ways. The correct response is to say that these models *do* form summary representations, but that they are summary representations of a special, flexible sort.

Let me add a word about the limitations of the models considered above. As McClelland and Rumelhart are the first to admit, many of these models have one major fault: they rely on a fixed set of representational primitives. Thus, our dog recognizer may have units interpreted as standing for size, color, and age, or whatever. But whatever the list, these are then the *only* dog features to which the network can be sensitive. If two dogs are different in a way not capturable by the list, the system cannot learn the difference. Indeed, the limitation is even worse than this. For if the linear combination of the values of the set of units excluding the one receiving external stimulation cannot be used to predict uniquely the activation of the currently, externally stimulated unit, the delta rule cannot guarantee perfect learning (see, e.g., McClelland, Rumelhart, and the PDP Research Group 1986, vol. 2, p. 181). It turns out, however, that this limitation can be overcome by having hidden units. These take no inputs from *outside* the system and send no outputs outside the system. Instead, they lie between the input and output units and can be used to mediate internal connectivity patterns so as to increase the number of input-output patterns the overall system can generate. In effect, the presence of hidden units allows the system to generate *new* representational primitives if these are needed to capture a pattern in the input. This greatly increases the power of the system. Using the powerful learning algorithms of recent connectionism (e.g., the generalized delta rule and the Boltzmann learning rule) these systems can generate the kinds of representation needed to solve problems that simple perceptron systems are provably unable to cope with.

The standard example of the use of hidden units is the exclusive "or" problem. Exclusive "or" is true just in case either *A* or *B* is true but not both *A* and *B* are true. You can have your cake or eat it, but not both. Now imagine a network whose input units are sensitive only to the presence or absence of *A* and the presence or absence of *B*. It would be easy if the problem was to get a network to reason using *inclusive* "or" (i.e., allow that "*A* or *B*" is true even if *both A* and *B* are true). Make one input unit pass a value of 1 when it sees *A*, and make another input unit pass a value of 1 when it sees *B*. Connect these to an output unit with a firing threshold of 1. If the system is given input of *A*, *B*, or *A* and *B*, the output unit will fire. Failing that, it won't. (see figure 5.6.) But this fails to cope with exclusive "or," since if the system is given input of *A* and *B*, the output unit will *still*

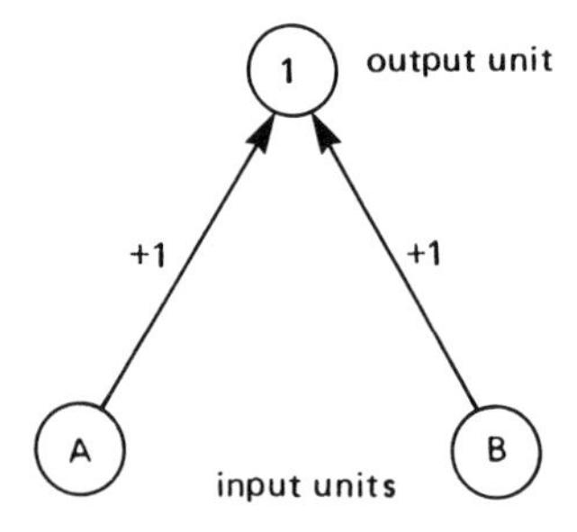

Figure 5.6
An inclusive *or* network

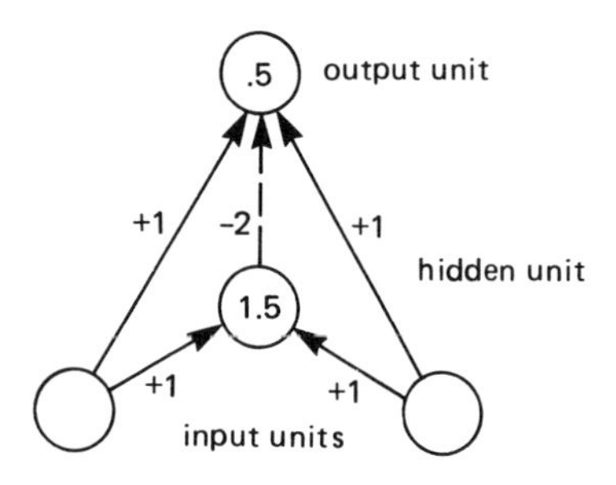

Figure 5.7
A simple exclusive *or* network with one hidden unit. From Rumelhart, Hinton, and Williams 1986, 321.

fire. If the input units are restricted to an *A* recognizer and a *B* recognizer, the solution is to have a hidden unit that fires just in case *A* and *B* recognizers *both* fire and that then passes a strongly inhibiting value to the output unit. The hidden unit thus represents the conjunction *A* and *B*, a representation that the system needs to solve the problem to which its two input units alone were unable to respond. Figure 5.7 shows a network that represents exclusive "or" (after McClelland, Rumelhart, and the PDP Research Group 1986, vol. 1, p. 321).

A major achievement of recent connectionism is its finding learning rules that, in the majority of cases, enable a system with hidden units to *learn* to deploy these in whatever ways are necessary to do justice to the structure of the input-output pattern required.[5] The first limitation of the distributed model of memory (its reliance on a fixed set of representational primitives) is thus surmountable, at least in theory, using only PDP apparatus. The second limitation is not so obviously surmountable in the same way. It is what I shall call the "serial control problem." Quite simply, the model "specifies the internal workings of some of the components of information processing and memory, in particular those concerned with the retrieval and use of prior experience. The model does not specify in and of itself how these acts of retrieval are *planned, sequenced* and *organised* into coherent patterns of behavior" (McClelland, Rumlhart, and the PDP Research Group 1986, vol. 2, p. 173, my emphasis). This problem will return in various guises to haunt us later on.

6 Biology Revisited

Chapter 4 detailed some biological constraints on natural intelligence. The PDP approach, it seems to me, goes a long way toward meeting these constraints. (More philosophical or conceptual constraints, which must be simultaneously satisfied, are dealt with in succeeding chapters.)

Natural intelligent systems, I argued, will set high store on robustness, fast sensory processing and motor control, and flexibility in dealing with new situations. Standard evolutionary pressures will favor economical, approximate solutions to problems. And the gradualistic holism of the evolutionary process means that the general form of solutions to early information-processing problems will heavily constrain the form of more recent ones. On all these counts, it appears, the PDP approach scores quite well. Let's review the situation.

PDP approaches deploy a means of encoding and processing information that is particularly well suited to evolutionarily basic tasks like low-level vision and sensorimotor control. For these are paradigm cases of tasks requiring the simultaneous satisfaction of a large number of soft constraints. Most of the basic tasks listed in chapter 4, section 4, fall into this category. PDP architectures (i.e., networks of units connected by excitatory and inhibiting links and set up to follow learning rules) would be a natural evolutionary choice as a means of performing such tasks at an acceptably fast speed. The use of such architectures as the basic form of computational solutions to more recent demands (e.g., semantic memory and even sentence processing—see the next chapter) thus respects the requirement of continuity with the form of earlier solutions.

Such an approach also brings a range of further benefits in the form of various emergent properties associated with parallel distributed means of encoding and retrieval. Thus, achievements that look from an intuitive mind's-eye perspective to be a number of *distinct* achievements, each in need of its own computational model, can be seen merely as different high-level ways of describing the same underlying form of processing—roughly, a constant attempt to configure the system to a best match for various exogenous and endogenous inputs. Fix the mode of encoding, updating, and retrieval and you get various forms of generalization, graceful degradation, default assignment, prototype extraction effects, and so forth. It begins to look very much as if PDP goes at least some way to finding the domes and arches out of which the spandrels visible to the mind's eye emerge. Many of the distinctions drawn at the upper level may still reflect our proper epistemological interests *without* marking any differences in the underlying computational form or activity of the system. Thus, we distinguish, say, memory and confabulation. But both may involve only the same pattern completing capacity. The difference lies, not in the computational substructure, but in the relation of that structure to states of the world that impinged on the knower. (For a discussion of how PDP from an *internal* perspective blurs the distinction between memory and confabulation, see McClelland, Rumelhart, and the PDP Research Group 1986, vol. 1, p. 80–81.)

Obviously, there is considerable and biologically attractive economy about all this. The use of one underlying algorithmic form to achieve so many high-level goals and the superpositional storage of data should appeal to a thrifty Mother Nature. And the double robustness of such systems (being capable of acting sensibly on partially incorrect data and capable of surviving some hardware damage) is a definite natural asset. Finally, the sheer *flexibility* of the system's use of its stored knowledge constitutes a major biological advantage (and philosophical advantage too, as we shall see). The capacity sensibly to deploy stored knowledge in a way highly sensitive to particular (perhaps novel) current needs will surely be a hallmark of any genuine intelligence.

In sum, the kind of approach detailed in the present chapter seems to offer a real alternative (or complement) to MIND-style theorizing. What makes PDP approaches biologically attractive is not merely their neurophysiological plausibility, as many seem to believe. They also begin to meet a series of more general constraints on any biologically and evolutionarily plausible model of human intelligence. PDP approaches may not be the *only* kind of computational approach capable of meeting such constraints. But work within the tradition of semantically transparent systems has so far failed to do so and instead has produced fragile, inflexible systems of uncertain neural plausibility.

Chapter 6
Informational Holism

1 *In Praise of Indiscretion*

Discretion, or at any rate discreteness, is not always a virtue. In the previous chapter we saw how a profoundly nondiscrete, parallel distributed encoding of room knowledge provided for rich and flexible behavior. The system supported a finely gradated set of potential, emergent schemata. In a very real sense, that system had no single or literal idea of the meaning of, say, "bedroom." Instead, its ideas about what a bedroom is are inextricably tied up with its ideas about rooms and contents in general and with the particular context in which it is prompted to generate a bedroom schema. The upshot of this is the double-edged capacity to shade or alter its representation of bedrooms along a whole continuum of uses informed by the rest of its knowledge (this double-edged capacity is attractive yet occasionally problematic—see chapter 9). This feature, which I shall call the "informational holism" of PDP, constitutes a major qualitative difference between PDP and more conventional approaches. This difference is tightly linked to the way in which PDP systems will fail to be *semantically transparent,* in the sense outlined in chapter one.

The present chapter has two goals. First, to elaborate on the nature of informational holism in PDP. Second, to discuss what *conceptual* relation parallel distributed encoding bears to the more general phenomenon of informational holism. On the latter issue, options range from the very weak (PDP sustains such holism, but so can work in the STS paradigm) to the very strong (only PDP can support such holism). The truth, as ever, seems to lie somewhere in between.

2 *Informational Holism in a Model of Sentence Processing*

Before we begin, it is worth reiterating an earlier warning. In what follows in this section, do not take any talk of what so-and-so *means* to a network too seriously, likewise with talk of what the network *knows,* and so on. This talk, e.g., of a network's shading its grasp of the meaning of a concept is a legitimate shorthand for claims of the following nature: the shading cf meaning found in natural-language understanding may be supported by the action of mechanisms that encode and retrieve data in the holistic PDP fashion. PDP researchers study such mechanisms using simple networks not in causal commerce with the referents of words like "bedroom" and not located in complex agents sharing a communal language. As such it is quite reasonable to withhold ascription of any actual grasp of meaning whatsoever to such networks. Still, if it is the action in us of something operating according to the principles of such networks that enables us to be as flexible and holistic in our grasp of meaning as we are, then the study of such mechanisms surely illuminates how we succeed in grasping the meanings we do. And this, rather than any deep philosophical confusion, is the basis of such talk. So much for the obvious but essential disclaimer.

To get a better idea of the shading power that constitutes the informational holism of PDP approaches, let us consider McClelland and Kawamoto's model of sentence processing (fully described in McClelland, Rumelhart, and the PDP Research Group 1986, vol. 2, chapter 19). The model aims to show in a highly simplified system the general way in which the capacity of PDP models simultaneously to satisfy multiple constraints might be exploited in a sentence processor. In particular, the model focuses on what is known as *case-role assignment*. Case-role assignment involves the process of deciding, among other things, which bit of the sentence specifies an agent, who does things; which bit specifies a patient, which gets things done to it; and which bit (if any) specifies the instrument of the doing. It turns out that a variety of factors combine to guide our interpretations: constraints of context, syntax, word-order, and basic semantics. The model developed is a fairly standard, distributed model of the kind described in the last chapter. It takes as input the result of a surface parse on the sentence and yields a reformulation as a set of semantic microfeatures. And it learns case-role assignment from paired presentations of such canonical input and desired output of case-role assignments. The kinds of microfeature used in the canonical format are unimportant in detail. In fact, the authors of the model used such features as volume, pointiness, breakability, and softness and allowed these features to adopt various values (see table 6.1). All this must be treated as a gross simplification. In the long run we might

Table 6.1
Microfeatures of a model of case-role assignment

Feature	Values
volume	small, medium, large
pointiness	pointed, rounded
breakability	fragile, unbreakable
softness	soft, hard

expect a reasonable model of the processes underlying grasp of sentence meaning to deploy a set of microfeatures that correspond to minute details of the visual, tactile, functional, and even emotive dimensions of human sensitivity to the world.

But even as it stands, the model exhibits some fascinating behavior whose extension to much more fine-grained kinds of microfeatural representation is intuitively obvious. According to its creators, the model exhibits "an uncanny tendency to shade its representation of the constituents of a sentence in ways that are contextually appropriate ... without any explicit training to do so" (McClelland and Kawamoto 1986, 276). I shall suggest that this property, which comes for free with parallel distributed storage and retrieval (at least with all genuinely *distributed* approaches), allows PDP models to provide a mechanism well suited to supporting a variety of important semantic phenomena. Of all the interesting properties of such models, this one, I believe, most firmly fixes any conceptual or qualitative advantages that PDP might have over other approaches. And indeed, McClelland and Kawamoto themselves describe the capacity to represent "a huge palette of shades of meaning" as being "perhaps ... the paramount reason why the distributed approach appeals to us" (1986, 314).

Here are some examples of the kind of shading they have in mind. The network was trained on various canonical reformulations of sentences involving, among other things, balls and breaking of objects. All the balls it learned about had the value "soft" along the dimension of softness. But all the objects it learned about as responsible for breaking things (hatchets, hammers, baseball bats, etc.) had the value "hard" along the dimension of softness. Because of the general tendency of PDP models to pick up on patterns in the input and to generalize to new cases, the subsequent behavior of the network should come as no surprise. Since all the objects used for breaking were hard, its knowledge of this range of other cases infects the way it deals with the sentence "the ball broke the vase." It correctly activates units for most of the features of the ball in the case role of instrument. But it diverges from the standard ball pattern along the softness dimension, activating "hard" instead of "soft." To us, this is a mistake. We know that soft balls are often responsible for breaking objects.

But the choice of the network to shade the meaning of ball in the context of breaking in the way it did cannot really be faulted. In light of its training, the use of the outlying information to the effect that all instruments of breaking are hard, not soft, seems quite insightful. As the authors put it, "As far as this model is concerned, balls that are used for breaking are hard, not soft" (McClelland and Kawamoto 1986, 305). This kind of shading of meaning in the light of the rest of the system's knowledge is the heart and soul of what I am calling the informational holism of distributed connectionist models. And it is bought by the simple fact that "different readings of the same word are just different patterns of activation [of microfeatures]; really different readings, ones that are totally unrelated, ... simply have very little in common. Readings that are nearly identical with just a shading of a difference are simply represented by nearly identical patterns of activation" (McClelland and Kawamoto 1986, 315).

The power of PDP systems to shade meanings across a whole continuum of cases enables them to model a number of effects. Most straightforwardly, it enables them to disambiguate words according to the context built up by the rest of the sentence. Thus take a sentence like "The bat ate the fruit." In this case the bat is clearly an animal not a cricket bat, and a PDP model could use the context of occurrence to determine this fact. The features that constitute the distributed representation of a live bat would be radically different from those appropriate to a cricket bat. The presence of the other words designating an act of eating fruit encourage activation of the features of the live bat.

This kind of effect, as McClelland and Kawamoto point out, can quite easily be captured in a conventional or in a *local* connectionist approach. In each case one would have a separate unit or memory store for each of the readings of "bat," and a set of rules or heuristics (in the conventional case) or a pattern of connectivity strengths (in the local connectionist version) determining which chunk to deploy when. This works until we need to model very-fine-grained differences in meaning. Separate chunks for a fruit bat and a cricket bat seem okay. But words seem to take on different shades of meaning in a continuously varying fashion, one that seems unspecifiable in advance. Thus, consider:

(1) The boy kicked the ball.
(2) The ball broke the window.
(3) He felt a ball in his stomach.

Sentences (1) and (2) are from McClelland and Kawamoto 1986, 315.) In case (1) we may imagine a soft, toy ball. In case (2) we imagine a hard ball (a tennis or cricket ball). In case (3) we have a metaphorical use: there is no ball in his stomach, but a feeling of a localized, hard lump. Everyday talk and comprehension is full of such shading effects according to overall context. Surely we don't want to commit ourselves to predetermining all such uses in advance and setting up a special chunk for the semantic meaning of each.

The PDP approach avoids such ontological excess by representing all these shades of meaning with various patterns in a single set of units representing microfeatures. The patterns for sentences (1) and (2) might share, e.g., the microfeature values spherical and game object, while the pattern for sentences (2) and (3) share the values small and hard. One interesting upshot here is the lack of any ultimate distinction between metaphorical and literal uses of language. There may be central uses of a word, and other uses may share less and less of the features of the central use. But there would be no firm, God-given line between literal and metaphorical meanings; the metaphorical cases would simply occupy far-flung corners of a semantic-state space. There would remain very real problems concerning how we latch on to just the *relevant* common features in understanding a metaphor. But it begins to look as if we might now avoid the kind of cognitivist model in which understanding metaphor is treated as the computation of a nonliteral meaning from a stored literal meaning according to high-level rules and heuristics. Metaphorical understanding, on the present model, is just a limiting case of the flexible, organic kind of understanding involved in normal sentence comprehension. It is not the icing on the cake. It is in the cake mix itself.[1]

So far, then, we have seen how the informational holism of distributed models enables them to support the representation of subtle gradations of meaning without needing to anticipate each such gradation in advance or dedicate separate chunks of memory to each reading. And we also saw hints of how this might undermine the rigidity of some standard linguistic categories like metaphorical versus literal use. One other interesting aspect of this informational holism concerns learning. The system learns by altering its connectivity strengths to enable it to re-create patterns in its inputs. At no stage in this process does it generate and store explicit rules stating how to go on. Superpositional storage of data means that as it learns about one thing, its knowledge of much else is automatically affected to a greater or lesser degree. In effect, it learns gradually and widely. It does not (to use the simplified model described) learn all about bats, then separately about balls, and so on. Rather it learns about all these things all at once, by example, without formulating explicit hypotheses. And what it can be said to know about each is informed by what it knows about the rest—recall the case of its knowledge that only *hard* things break things.

In sum, PDP approaches buy you a profoundly holistic mode of data storage and retrieval that supports the shading of meanings and allows gradual learning to occur without the constant generation and revising of explicit rules or hypotheses about the pattern of regularities in the domain.

3 *Symbolic Flexibility*

Smolensky (1987) usefully describes PDP models as working in what he calls the subsymbolic paradigm. In the subsymbolic paradigm, cognition is not modeled by the manipulation of machine states that neatly match (or stand for) our daily, symbolic descriptions of mental states and processes. Rather, these high-level descriptions (he cites goals, concepts, knowledge, perceptions, beliefs, schemata, inferences, actions) turn out to be useful labels that bear only approximate relations to the underlying computational structure. He argues that work in the subsymbolic (or distributed connectionist) paradigm aims to do justice to the "real data on human intelligent performance," i.e., to clinical and experimental results, while settling for merely emergent approximations to our high-level descriptive categories. The essential difference between the subsymbolic and the symbolic approach, as Smolensky paints it, concerns the question, Are the semantically interpretable entities the very same objects as those governed by the rules of computational manipulation that define the system?

In the symbolic paradigm, the answer is yes. Consider the STS approach we sketched way back in chapter 1. Here we find computational operations directly applied to high-level descriptions of mental states presented as a means of capturing the computational backdrop of mind. Thus, we might find a model of scientific discovery in which operations are performed on states directly interpretable as standing for particular hypotheses concerning the laws governing some data. Against this kind of approach, the subsymbolic theorist urges that the entities whose behavior is governed by the rules of computational manipulation that define the system need not share the semantics of the task description. For what is so governed is just the activation profiles of individual units in a network. And in a highly distributed model these units in the end will have no individual semantic interpretation, or at least none that maps neatly and projectibly onto our ordinary concepts of the entities to be treated in a model of the processing involved. Rather, what gets semantically interpreted will be general patterns of activation of such units. A single high-level concept like that of a kitchen or a ball will, we saw, be associated with a continum of activation patterns corresponding to the subtly different ideas about kitchen or ball that we entertain in various circumstances. Smolensky puts it nicely in the following passage:

> In the symbolic approach, symbols (atoms) are used to denote the semantically interpretable entities (concepts). These same symbols are the objects governed by symbol manipulations in the rules which define the system. The entities which are capable of being semantically interpreted are also the entities governed by the formal laws that define the system. In the subsymbolic paradigm, this is no longer true. The semantically interpreted entities are patterns of activation over a large number of units in the system, whereas the entities manipulated by formal rules are the individual activations of cells in the network. The rules take the form of activation passing rules, of essentially different character from symbol manipulation rules. (1987, 100)

The claim, in effect, is that PDP systems need not (and typically will not) be semantically transparent in the sense introduced in chapter 1 above. Such a claim may not seem immediately plausible for the following reason. A system will count as semantically transparent just in case the entities found in a top-level task analysis of what the system does have neat syntactic analogues whose behavior is governed by the computational rules (explicit *or* tacit) of the system. Now clearly, it will not do to say that just because individual units cannot be treated as the syntactic analogues of such entities (e.g., as "coffee," "ball," "kitchen," and so on) the condition fails to be met. For why not treat patterns of activation of such units as the required analogues? The behavior of such patterns surely *is* governed by the computational rules of the system.

This is where the requirement that such analogues be *projectible* comes in. Consider a sentence like "the ball broke the window." A conventional AI system dealing with such a sentence will have a syntactic analogue (first in, say, LISP and hence down to machine code) for "ball" and "window." Consider now a connectionist representation of the same sentence. There will be a pattern of active units, and it may well be possible to nonarbitrarily isolate a subset of that pattern that, we would like to say, stands for "ball". But that subpattern, it is important to note, will vary from context to context. "Ball" as it occurs in "The ball broke the window" will have a different (though doubtless partially overlapping) syntactic analogue to "ball" as it occurs in "The baby held the ball." In one case the hardness related microfeatures will be active. In the other case, not. Thus, although in each individual case we can isolate a connectionist syntactic analogue for the entities spoken of in a conceptual account, these entities will not be neatly projectible, i.e., the *same* syntactic entity will not continue to correlate in other cases with the top-level semantic entity. This is the real sense in which PDP systems can constitute a move away from semantic transparency.

The example given above could be multiplied. Smolensky (1988) makes similar comments about the symbol "coffee" as it occurs in various contexts. And we could say the same for "bedroom" as it occurs in the sofa-including and sofa-excluding contexts treated earlier. The general point, then, is that "the context [in PDP systems] alters the internal structure of the symbol: the activities of the sub-conceptual units that comprise the symbol—its subsymbols—change across contexts" (Smolensky 1988, 17). Smolensky formalizes this point as a characteristic of the highly distributed PDP systems he is interested in as follows: "In the symbolic paradigm the context of a symbol is manifest *around* it and consists of *other symbols*; in the subsymbolic paradigm the context of a symbol is manifest *inside* it, and consists of subsymbols" (Smolensky (1988, 17). Both the intrinsic holism and flexibility of PDP systems can be seen to flow from this fact.

4 Grades of Semantic Transparency

Using this apparatus as a base, Smolensky (1988) formulates an interesting picture of the cognitive terrain. He suggests that some human knowledge (e.g., public scientific knowledge) exists in the first instance as linguistic items such as the principle "energy is conserved." Human beings, he suggests, may use such knowledge by deploying a virtual machine adapted to manipulate analogues of such linguistic representations. Such explicitly formulated knowledge he calls "cultural knowledge." The "top-level conscious processor" of an individual is precisely, he argues, a virtual machine adapted to that end. This machine, which is realized by a PDP substructure, he calls the "conscious rule interpreter." It is contrasted with what he calls the "intuitive processor." The distinction again depends on the kind of entities processed. The conscious rule interpreter actually takes as its syntactic objects the semantic entities we use in describing the task domain (e.g., "energy"). The intuitive processor, by contrast, takes as its objects distributed microfeatural representations of the kind treated above. These representations, we saw, bear only a fluid and shifting relationship to the semantic entities (like "coffee" and "ball") spoken of at the conceptual level. It thus follows that the programs running on the conscious rule interpreter have a syntax and semantics comparable to our top-level articulation of the domain. (This is no accident: they are precisely models of that top-level articulation.) While the programs running on the intuitive processor do not. In my terminology, programs running on the conscious rule interpreter will be semantically transparent and the semantics will seep neatly down to the formal level, while those running on the intuitive processor will not. The intuitive processor is quite clearly to be seen as the more evolutionarily basic of the two and is reponsible (he says) for all animal behavior and much human behavior, including: "perception, practised motor behavior, fluent linguistic behaviour, intuition in problem solving and game-playing—in short, practically all of skilled performance" (Smolensky 1988, 5).

There need not, however, be an all-or-nothing divide between the semantically transparent processing of the conscious rule interpreter and the semantically opaque processing of the intuitive processor. For the cognitive system itself is presumed to be at root a subsymbolic system that, to a greater or lesser degree in various cases, *approximates* to the behavior of a symbolic system manipulating conceptual entities. The greater the so-called dimension shift between the conceptual description and the semantic interpretation of the units in the network, the rougher such an approximation becomes. Thus, the model of emergent schemata we examined earlier behaves in a wide variety of cases as if it had a standard, rigid schema of bedrooms. But it diverges in cases of nonstandard rooms (e.g., a bedroom with a sofa). The range of such divergence will increase with the distance between the conceptual entity ("kitchen" etc.) and the microfeatures in the network. In a treatment with many microfeatures, where such items as "bed" and "sofa" are merely approximate top-level labels for subtle and context-sensitive complexes of geometric and functional properties, the distance will be great indeed, and a conceptual model only a very high approximation.

Another route to the approximation claim is to regard the classical accounts as describing the *competence* of a system, i.e., its capacity to solve a certain range of well-posed problems (see Smolensky 1988, 19). In idealized conditions (sufficient input data, unlimited processing time) the PDP system will match the behavior specified by the competence theory (e.g., settling into a standard kitchen schema on being given "oven" and "ceiling" as input). But outside that idealised domain of well-posed problems and limitless processing time, the performance of a PDP system will diverge from the predictions of the competence theory in a pleasing way. It will give sensible responses even on receipt of degraded data or under severe time constraints. This is because although describable in that idealized case as satisfying hard constraints, the system may actually operate by satisfying a multitude of soft constraints. Smolensky here introduces an analogy with Newtonian mechanics. The physical world is a quantum system that looks Newtonian under certain conditions. Likewise with the cognitive system. It *looks* increasingly classical as we approach the level of conscious rule following. But in fact, according to Smolensky, it is a PDP system through and through.

In the same spirit Rumelhart and McClelland suggest: "It might be argued that conventional symbol processing models are macroscopic accounts, analogous to Newtonian mechanics, whereas our models offer more microscopic accounts, analogous to quantum theory.... Through a thorough understanding of the relationship between the Newtonian mechanics and quantum theory we can understand that the macroscopic level of description may be *only an approximation* to the more microscopic theory" (Rumelhart and McClelland 1986, 125). To illustrate this point, consider a simple example due to Paul Smolensky. Imagine that the cognitive task to be modeled involves answering qualitative questions on the behavior of a particular electrical circuit. (The restriction to a single circuit may appall classicists, although it is defended by Smolensky on the grounds that a small number of such representations may act as the chunks utilized in general purpose expertise—see Smolensky 1986, 241.) Given a description of the circuit, an expert can answer questions like "If we increase the resistance at a certain point, what effect will that have on the voltage, i.e., will the voltage increase, decrease, or remain the same?"

Suppose, as seems likely, that a high-level competence-theoretic specification of the information to be drawn on by an algorithm tailored to answer this question cites various laws of circuitry in its derivations (what

Smolensky refers to as the "hard laws" of circuitry: Ohm's law and Kirchoff s' law). For example, derivations involving Ohm's law would invoke the equation

voltage = current × resistance.

How does this description relate to the actual processing of the system? The model represents the state of the circuit by a pattern of activity over a set of feature units. These encode the qualitative changes found in the circuit variables in training instances. They encode whether the overall voltage falls, rises, or remains the same when the resistance at a certain point goes up. These feature units are connected to a set of what Smolensky calls "knowledge atoms," which represent patterns of activity across subsets of the feature units. These in fact encode the legal combinations of feature states allowed by the actual laws of circuitry. Thus, for example, "The system's knowledge of Ohm's law ... is distributed over the many knowledge atoms whose subpatterns encode the legal feature combinations for current, voltage and resistance" (Smolensky 1988, 19). In short, there is a subpattern for every legal combination of qualitative changes (65 subpatterns, or knowledge atoms, for the circuit in question).

At first sight, it might seem that the system is merely a units-and-connections implementation of a lookup table. But that is not so. In fact, connectionist networks act as lookup tables only when they are provided with an overabundance of hidden units and hence can simply memorize input-output pairings. By contrast, the system in question encodes what Smolensky terms "soft constraints," i.e., patterns of relations that usually obtain between the various feature units (microfeatures). It thus has general knowledge of qualitative relations among circuit microfeatures. But it does *not* have the general knowledge encapsulated in *hard* constraints like Ohm's law. The soft constraints are two-way connections between feature units and knowledge atoms, which *incline* the network one way or another but do not *compel* it, that is, they can be overwhelmed by the activity of other units (that's why they are soft). And as in all connectionist networks, the system computes by trying simultaneously to satisfy as many of these soft constraints as it can. To see that it is not a mere lookup tree of legal combinations, we need only note that it is capable of giving sensible answers to (inconsistent or incomplete) questions that have no answer in a simple lookup table of legal combinations.

The soft constraints are numerically encoded as weighted inter-unit connection strengths. Problem solving is thus achieved by "a series of many node updates, each of which is a *microdecision* based on formal *numerical* rules and numerical computations" (Smolensky 1986, 246).

The network has two properties of special interest to us. First, it can be shown that if it is given a well-posed problem and unlimited processing time, it will always give the correct answer as predicted by the hard laws of circuitry. But, as already remarked, it is by no means bound by such laws. Give it an ill-posed or inconsistent problem, and it will satisfy as many as it can of the soft constraints (which are all it really knows about). Thus, "outside of the idealised domain of well-posed problems and unlimited processing time, the system gives sensible performance" (Smolensky 1988, 19). The hard rules (Ohm's law, etc.) can thus be viewed as an external theorist's characterization of an idealized subset of its actual performance (it is no accident if this brings to mind Dennett's claims about the intentional stance—see Dennett 1981).

Second, the network exhibits interesting *serial* behavior as it repeatedly tries to satisfy all the soft constraints. This serial behavior is characterized by Smolensky as a set of *macrodecisions* each of which amounts to a "commitment of part of the network to a portion of the solution." These macrodecisions, Smolensky notes, are "approximately like the firing of production rules. In fact, these 'productions' 'fire' in essentially the same order as in a symbolic forward-chaining inference system" (Smolensky 1988, 19). Thus, the network will look as if it is sensitive to hard, symbolic rules at quite a fine grain of description. It will not *simply* solve the problem "in extension" as if it knew hard rules. Even the *stages* of problem solving may look as if they are caused by the system's running a processing analogue of the steps in the symbolic derivations available in the competence theory.

But the appearance is an illusion. The system has no knowledge of the objects mentioned in the hard rules. For example, there is no neat subpattern of units that can be seen to stand for the general idea of resistance, which figures in Ohm's law. Instead, some sets of units stand for resistance at R_1, and other sets for resistance at R_2. In more complex networks the coalitions of units that, when active, stand in for a top-level concept like resistance are, as we saw, highly context-sensitive. That is, they vary according to context of occurrence. Thus, to use Smolensky's own example, the representation of coffee in such a network would not consist of a single recurrent syntactic item but a coalition of smaller items (microfeatures) that shift according to context. Coffee in the context of a cup may be represented by a coalition that includes the features (liquid) and (contacting-procelain). Coffee in the context of jar may include the features (granule) and (contacting-glass). There is thus only an "approximate equivalence of the 'coffee vectors' across contexts," unlike the "exact equivalence of the coffee tokens across different contexts in a symbolic processing system" (Smolensky 1988, 17). By thus replacing the conceptual symbol "coffee" with a shifting coalition of microfeatures, the so-called dimension shift, such systems deprive themselves of the structured mental representations deployed in both a classical competence theory and a classical symbol-processing account (level 2). Likewise, in the simple network described, there is no stable representation that stands for resistance (just as in the famous past-tense network there is no stable, recurrent entity that stands for verb stems [see chapter 9]).

It seems, then, that by treating subsymbolically the entities spoken of in our conceptual-level descriptions, we buy the flexibility, shading, and general lack of rigidity and brittleness *required* of a system if its subsequent behavior is ever to warrant the ascription to it of a genuine grasp of concepts. Symbolic flexibility of understanding is brought about by the increased low-level variability of the PDP approach. In this way such systems may avoid the excessive rigidity and lack of insight in conventional AI so thoroughly bemoaned by Dreyfus (see chapter 2). Notice also that the subsymbolic model remains formal in the sense outlined in chapters 1 and 2. It is a microfunctionalist theory as defined in chapter 2, section 6. That is, it specifies a system only in terms of input-output profiles for individual units and thus is not crucially dependent on any particular biological substrate. But the entities figuring in the formal profile do not correspond to, or otherwise nearly preserve, the boundaries of any conceptual-level description of thought. This is all good news, especially in the light of our strictures against projecting conscious conceptual categories back into the head (see chapter 3). But we must now consider a slightly tricky question concerning the *nature* of the relation between PDP substructures and the various desirable symbolic properties they seem capable of supporting.

5 *Underpinning Symbolic Flexibility*

PDP approaches, we saw, are well suited to modeling the kind of symbolic flexibility associated with human understanding. But what *kind* of relation does the locution "well suited" so comfortingly gloss? Here are some possibilities.

(1) Only a PDP style substructure can support the kind of flexible understanding required (uniqueness).

(2) Maybe non-PDP systems can do the job. But semantically transparent approaches simply *cannot* achieve such flexibility (qualified liberalism).

(3) There is nothing special about a PDP capacity to support flexible understanding (unqualified liberalism).

Just which option we choose must depend, in part, on what we understand by "a PDP approach." We could mean:

(a) A model based firmly on one of the current algorithmic forms deployed by McClelland, Rumelhart, and the PDP Research Group (e.g., a model using a Boltzmann learning algorithm or the generalized delta rule).

(b) A model that is *formally specified* by a description of simple units in parallel, excitatory and inhibitory connections and by value-passing rules of one kind or another, and for which a semantic interpretation of that formal level of specification involves a dimensional shift away from the categories and entities involved at the ordinary level of task analysis. (Call this "architectural PDP.")

(c) A model (any model) that supports the special qualities of flexible retrieval, holistic storage and learning, and context-sensitive shading of meaning. (Call this "functional PDP.")

The (a) reading is surely far too strong. No one seriously believes that the algorithms currently studied within PDP tell the final connectionist story about neural computation. And a (c) reading seems too vague. There seems no good reason to expect the label "parallel distributed processing" to be naturally applicable to just *any* system capable of supporting flexibility and holism of the relevant kind. This leaves us with the (b) reading, architectural PDP. "A PDP approach" means an approach using formal, network models whose semantic interpretation as a network of simple units and value-passing rules involves a dimensional shift away from top-level conceptual entities and down to more fine-grained, microfeatures. This is the reading I adopt throughout this book.

What, then, of our earlier options? The uniqueness claim (option 1) seems impossible to support. Since we have no idea of the range of computational strategies that might remain to be discovered, no one can say in advance that *only* a PDP approach can achieve some goal. The nearest we get to any such argument is the following. Suppose you are restricted to generating computational models that are realizable in the kind of biological hardware that humans possess. Because of what we know of the speed of such hardware, it looks as if some tasks (primarily the perceptual completion and interpretation tasks discussed in chapter 5) could be performed as fast as they are *only if* some kind of parallelism is involved. Our current focus, however, is on the kind of flexible understanding of meaning embodied in contextual-shading effects. And the problem domain here is by no means as well understood as that of say, low-level vision. So as yet it would be foolish even to endorse a qualified version of option 1, i.e., one reducing the uniqueness claim to "unique if the system must be implemented on a neural network."

This observation, however, should not be used as an argument for unqualified liberalism (option 3). For it is by no means clear that the kind of flexible deployment of knowledge and continuous gradations of meaning found in PDP are fully available to semantically transparent systems, i.e., systems in which projectible syntactic entities act as analogues for the entities invoked in a top-level analysis of a task and are manipulated by the computational operations that define the system. For the flexible, context-sensitive retrieval and deeply holistic storage of data found in PDP approaches is a direct function of the way such approaches treat top-level conceptual entities, e.g., building "ball" (section 3 above) or "coffee" (Smolensky 1988) out of microfeature complexes whose precise constitution varies according to the details of a particular context of occurrence.

To make these points vivid, recall the example of the network given sentences involving balls and breakage. The network looked as if it had made the following inference: if most objects used for breaking things are hard and a ball is used for breaking things, contradict the central case and represent the ball as being hard. We can imagine this inference formalized into a general rule. If most *f*s are *g* and *x* is an *f*, then *x* is to be represented as having the feature *g*. All well and good. In some cases, at least, a semantically transparent model following a rule like this would exhibit the kind of shading of meaning we require. But note the following. First, we need to specify what we mean by "most." So our flexible system will get a little stiff around the edges. A PDP system with continuous strengths of activation of microfeature units can avoid this stiffness. Second, we can depend on such tactics and stay faithful to a semantically transparent methodology *only* if the softening rules specify operations on features that are likely elements of top-level reflection on a task (e.g., the feature "hard"). If we desire even finer gradations of meaning, we will need sets of microfeatures for these features also, until ultimately the individual units won't be semantically transparent at all. Recall also (chapter 5) that a feature of some PDP systems is that they generate their *own* microfeature-detector units as required. Differences in *patterns* of activation of these self-taught supermicrofeature units may underpin very fine differences in meaning. This deep flexibility seems unavailable if we insist that our computational operations apply to semantically interpretable entities of a kind we pick out in advance. Third, even if we put this worry aside, the sheer proliferation of rules and heuristics needed to even approximate human flexibility and shades of understanding will be enormous (recall Dreyfus's observations reported in Chapter 2).

Now suppose we manage in a serial (but not semantically transparent) system to simulate the required range of flexibility, shading potential, etc. *Even then* there is a suspicion that the simulated holism thus achieved will never be as *deep* as the natural holism of a PDP system. For there will always remain conceptually important differences in the counterfactual behavior of the systems. Thus, suppose you have a serial model with a discrete representation of a square and a discrete representation of a rectangle. And we add a rule that says that squares inherit the properties of rectangles. In a PDP system this might be achieved by having a single network with a distributed representation of square and rectangle stored superpositionally over a set of microfeature units that represent, e.g., that the sides meet at ninety-degree angles. We simply *could not* reach into such a system and delete its knowledge of the ninety-degree feature of squares without *also* affecting its knowledge of the ninety-degree feature of rectangles. Contrast the artificial "holism" of the serial model. Here, there is a possibility of deleting the *rule* for knowledge of squares while leaving the knowledge of rectangles untouched. Of course, a serial approach may well have other resources at its disposal. The point is that the holism of the PDP model is an *ineliminable* feature of its mode of data storage and is thereby robust across counterfactual cases, which may shatter the more fragile holism of a more conventional version.

6 PDP and the Nature of Intelligence

There is a standard response to any attempt to establish a qualitative difference between PDP and conventional approaches. It relies on the supposed portability of software. Parallel processes can always be simulated in serial fashion if we are either unconcerned with how long they take to run or willing to imagine arbitrarily fast serial machines (possibly exceeding the speed of light itself). This fact gives rise to the following kind of remark: "A program running on a parallel machine that produced some sort of intelligence will also run on a serial machine, *and this is enough to show the hardware irrelevant for explaining the nature, if not the evolution, of that particular intelligence*" (Krellenstein 1987, 155; my emphasis). This kind of argument, I think, underlies much of the resistance to PDP approaches encountered in mainstream cognitive science. But the issues involved are by no means as cut and dried as Krellenstein's bald statement leads us to expect. Granted, the mere use of parallel hardware is hardly going to be the decisive factor in the explanation of some particular intelligence. Indeed, if we ignore time and speed constraints, anything a real PDP machine can run could be run on a serial simulation of the machine, that is, on a *virtual* PDP machine created on a serial processor (e.g., the Symbolics LISP Machine). Of course, time and speed may not be irrelevant features when it comes to assessing the intelligence of a system. We do not allow that someone who takes hours to solve a problem by the serial enumeration of all the possibilities has thereby demonstrated the same intelligence as someone who solves it quickly by a highly selective and well-chosen search procedure. As we saw in chapter four, time and the efficient use of resources are at a premium for natural intelligence (and where else is our idea of intelligence to be grounded?). Thus, it is possible to dispute the claim that, "if the program runs too slowly on the serial machine to be useful we would not say that it no longer demonstrates intelligence but only that it is too slow, or that the particular approach, though successful, is impractical" (Krellenstein 1987, 155).

However, I shall not pursue this line. Instead, for the purpose of argument I shall accept the point about irrelevance of hardware. For much of the interest of PDP, it seems to me, lies *not* in the notion of parallel hardware so much as in the particular properties of *any* virtual machine (whatever its hardware base) running parallel cooperative algorithms of the kind we have been discussing. The particular properties I have in mind (some of which we have already met) can be grouped under the headings of "Searches" and "Representation."

Searches

Even the most conventional work in AI accepts a relationship between degree of intelligence and efficiency of search. The point of heuristic searches (chapter 1, section 4) is precisely to increase the *intelligence* of the system by reducing the extent of the search space it must traverse to solve a particular problem. Some problems require finding the best way of simultaneously satisfying a large number of soft constraints. In such cases PDP, we saw, provides an elegant solution to the problem of efficient, content-sensitive searches. A search is driven by partial descriptions and involves simultaneously applying competing hypotheses against one another. It is not dependant on fully accurate or even consistent information in the partial description. PDP methods thus introduce a qualitatively different kind of search into AI. This kind of search may not *always* be the best or most efficient option; as always, it all depends on the nature of the space involved.

Representation

We have seen how PDP models (or those PDP models which interest us) employ distributed representations in superpositional storage. The deep informational holism of PDP depends on this fact. Besides explaining shadings of meaning, discussed above, this mode of representation has important side effects. Some of these were discussed in chapter 5 (e.g., graceful degradation, generalization, content-addressible retrieval). One very important side effect that I have not yet fully discussed is cross talk.

Cross talk is a distinctive PDP pathology that occurs when a single network processes multiple patterns with common features. Because the patterns have common features, there is a tendency for the overall pattern of activation appropriate to one pattern to suffer interference from the other pattern because the system tries simultaneously to complete the other pattern. At times I have praised this property by calling it "free generalization." When it occurs in some contexts, however, it is a source of errors. Thus, a system that seeks to recognize words might occasionally fall into error when two simultaneously presented words share some features, e.g., "sand" and "lane" might get read as "land" and "sane" (see McClelland 1986, 139). Or in a model of skilled typing, the keystrokes appropriate to two successive letters may interfere and cause overlap or exchange (see Rumelhart and Norman 1982).

Cross talk might be regarded as an irritating side effect of parallel distributed representation were it not for two facts. First, the *kind* of error caused by cross talk is familiar to us in many areas of life. (Recall once again the difficulty of remembering the phone numbers of two friends if the numbers are very similar. We tend to mix up elements of each. That's cross talk.) It turns out that rather fine-grained experimental data on human performance, including such error profiles, can be explained and even predicted using PDP models (see McClelland 1986, 139–140). This is presumably good news for PDP considered as an attempt to model human processing mechanisms.

But second and more important, the source of such occasional errors is *also* the source of much of the power and flexibility of such systems. The tendency to generalize is one example of this. Another even more powerful effect is to provide a computational underpinning for the capacity to think analogically, to see parts of our knowledge as exemplifying patterns and structures familiar to us from other parts. This tendency to see one part of an experience in terms of another, to see links and similarities and patterns all around us, is surely at the heart of human creativity. It is what Hofstadter (1985) calls "fluid thinking."

At its best cross talk may underpin such classic anecdotes as Kekule's discovery of the benzine ring after dreaming of snakes biting their own tails, or Wittgenstein's capacity to see a court room simulation of a motor accident using model vehicles as indicating a deep truth about the nature of language. At its worst, cross talk pathologically amplified may explain the schizophrenic's tendency to see deep links between every item of daily experience. Either way, the idea of an informational substructure that forces us continuously to seek and extend patterns (the very substructure responsible for cross talk) is psychologically highly attractive. For, as Hofstadter puts it:

> People perceive patterns anywhere and everywhere, without knowing in advance where to look. People learn automatically in all aspects of life. These are just facets of common-sense. Common-sense is not an "area of expertise," but a general—that is, domain-independent—capacity that has to do with fluidity in representation of concepts, an ability to sift what is important from what is not, an ability to find unanticipated analogical similarities between totally different concepts. (1985, 640).

Nobody, Hofstadter included, claims that current PDP systems provide good models (or even hints) of how we come to exhibit all these properties. Especially difficult and important is the problem of how we fix on the important similarities between two concepts or domains and ignore the unimportant ones. Nonetheless, cross talk does seem to suggest the beginnings of a partial answer to the more general question of how we can find *unanticipated* similarities, and, as we saw, PDP also provides some account of the fluidity of our representations of concepts. These are features that depend crucially on the *architecture* of the PDP approach, that is, on the (possibly virtual) setup of units, output functions, patterns of connectivity, propagation rules for spreading activation, and learning rules. In this sense, some of the most interesting qualitative advantages of PDP models are gained at a level much closer to the implementation detail (though not yet the hardware level) than is standardly considered important. That is, they are a direct function of the capacity of this architecture to support distributed, superpositionally stored representations.

Now, it may be that other, as yet undiscovered architectures have the same property, and if so, the uniqueness claim (section 4, option 1) would have to be whole-heartedly rejected. But it certainly does not look as if a semantically transparent approach to programming will be able elegantly to support the desired properties. At the end of the day, therefore, a qualified liberalism (section 4, option 2) seems to be warranted. Some may choose to see PDP as just suggesting new ways of implementing conventional AI approaches, like scripts and frames. But by being rendered in the PDP distributed, superpositional style, these approaches gain considerably in power and flexibility. Indeed, they gain flexibility to such an extent that it often seems sensible to view the standard constructs as Smolensky does, as useful approximations to the fine-grained picture presented by PDP. Either way, it remains true to say that the use of a PDP architecture opens up new and qualitatively different avenues of searches and representation to those so far explored in conventional AI.

7 *An Equivalence Class of Algorithms*

PDP would have maximum philosophical significance just in case the following result could be established. The result may be put like this:

> There exists a formal specification that picks out algorithmic forms that are instances of a PDP model of one kind or another, and for conceptual reasons it is clear that any system capable of exhibiting the kind of rich and flexible behavior that warrants talk of that system as knowing, meaning, understanding, or having thoughts must rely on one of those algorithmic forms.

Call this "the equivalence-class conjecture." Roughly, it states that PDP descriptions could in principle be used to specify an equivalence class of algorithmic forms, and deploying some member of that class is a constitutive requirement of bona fide thinking. It will be seen that this conjecture consists essentially in a combination of the uniqueness claim (rejected earlier) and the claim that the kind of flexible, context-sensitive response made possible by a PDP architecture is essential to thinking.

The latter claim, it seems to me, is well founded. Many philosophical worries about creating artificial intelligence were based precisely on an observed lack of flexible, commonsense deployment of stored information and on a lack of fluid, analogical reasoning. (Recall the discussion of Dreyfus in chapter 2, and see Hofstadter 1985.) Likewise, the capacity to shade the meaning of various high level concepts and the tendency to learn concepts in large, mutually dependent groups may reasonably be seen as part of the very idea of grasping a concept. The features that PDP work seems capable of supporting in time may thus be philosophically significant parts of an account of understanding.

It might be argued that this is to give our notion of understanding an overly anthropocentric twist. Humans and higher animals may well exhibit the features mentioned. But why should it be required of *all* understanders that they conform to our model? To this line of thought I can only reply that the concept of understanding was formed to describe the behavior of humans and higher animals. It need not be indefensibly anthropocentric to believe that certain features of these uses must carry over to any case where the concept is properly used.

Suppose, then, that we agree that essential to understanding are some of the features that a PDP approach looks well suited to underpinning. For its truth the equivalence-class conjecture would then require, that we find conceptual reasons why *only* members of the PDP class of models can support such features. It is this part of the conjecture that looks insupportable. Conceptual ties may be felt to exist between, say, the use of distributed representations with a microfeature semantics and the kind of deep holism and flexible understanding we require. But this alone is insufficient to count as a PDP model according to our definition (section 4 above). What we cannot rule out in advance is the possibility of some as yet undiscovered but distinctly non-PDP architecture supporting distributed representations and superpositional storage and hence providing all the required features in a new way. If this were to prove possible, we would be in the midway position described in chapter 3, section 7, that seems to blur the divide between the constitutive and the causal. The dream that PDP can specify a class of algorithms essential to thinking is precisely the dream of finding a formal unity among the set of substructures capable of supporting thought. Of course, it may turn out that the only physically possible way of achieving distributed representation and superpositional storage to the required degree involves the use of a (possibly virtual) value-passing network. If we could see just why this should be so (as a result of, say, physical limitations on processing imposed by the speed of light), we would have quite a strong result. The PDP substructure would be revealed as a naturally necessary condition for the flexible behavior (including linguistic behavior) that is conceptually essential to the ascription of thoughts.

Briefly, the various possibilities look like this.

- The weakest interesting claim is that a PDP substructure supports flexible understanding and behavior, and these are essential for an ascription of understanding.
- The intermediate claim is that a PDP substructure is naturally necessary for flexible understanding and behavior, and these are essential for an ascription of understanding.
- The strong claim is that a PDP substructure is required on conceptual grounds (i.e., independently of physical limitations like the speed of light) for flexible understanding and behavior, and these are essential for an ascription of understanding.

I believe it is too early to try to force a choice between these claims. As we saw, it seems that the strongest claim presently eludes us. And even the intermediate claim is suspect. That said, I believe philosophical interest still attaches to the PDP appraoch, since it begins to demonstrate how a physical system is at least *able* to support various features that *must* be exhibited by any system that warrants description in a mentalistic vocabulary. There are those whose idea of philosophical interest would place such a result outside the sphere of their proper concerns. For such philosophers nothing short of the truth of the equivalence-class conjecture would motivate a claim of philosophical significance. Here we must simply differ and be content to get as clear as we can both about the possible relations between PDP models and thoughts and about the grounds for expecting any particular relation to hold.[2]

Chapter 7
The Multiplicity of Mind: A Limited Defence of Classical Cognitivism

1 *Of Clouds and Classical Cognitivism*

Every silver lining has a cloud, and PDP is no exception. There are classes of problems to which PDP approaches are apparently ill suited. These include the serial-reasoning tasks of logical inference, the temporal-reasoning tasks of conscious planning, and perhaps the systematic-generative tasks of language production. PDP seems to be nature's gift to pattern recognition tasks, low-level vision, and motor control.[1] But as we proceed to higher, more-abstract tasks, the PDP approach becomes less and less easy to employ. This is, of course, just what we would expect on the basis of our earlier conjectures. A PDP architecture may have been selected to facilitate carrying out evolutionarily basic tasks involving multiple simultaneous satisfaction of soft constraints. Vision and sensorimotor control are prime examples of such tasks. Other tasks—especially the relatively recent human achievements that classic cognitivism focuses on—involve complex sequential operations that may require a system to follow explicit rules. Conscious reasoning about chess playing, logic, and conscious attempts to learn to drive a car are examples of such tasks. Where the conscious-reasoning aspects of such tasks are concerned, the standard architecture of classical-cognitivist models offers an excellent, design-oriented aid to their solution. In these models an explicitly programmed CPU (central processing unit) performs sequential operations on symbolic items lifted out of memory. The architecture is perfectly suited to the sequential application of explicit rules to an ordered series of symbol strings.

Such sequential, rule-following acrobatics are not the forte of PDP. They may not be beyond the reach of a PDP approach, but they certainly do not come naturally to it. It may be significant to notice, however, that these sequential rule-following tasks aren't our forte either. They are the tasks, that human beings find hardest, the ones we tend to fail at. And often, after we cease to find them hard (after we are good at chess or logic or driving a car), we also cease to have the phenomenal experience of consciously and sequentially following rules as we perform them.

The thought I develop in this chapter concerns a possible multiplicity of virtual cognitive *architectures*. The idea is that for some aspects of some reasoning tasks, we might be forced to emulate a quite different kind of computing machine. For example, to perform conscious deductive reasoning, we might emulate the architecture of a serial Von Neumann machine. This idea is by no means new, but its full significance is not generally appreciated. The picture of multiple architectures forms half of a partial defence of classical cognitivism. (For the other half, see the appendix, in which I provide an independent argument for a multiplicity of styles of cognitive *explanation*, a multiplicity demanded *whatever* the architectural facts are.)

2 *Against Uniformity*

All too often, the debate between the proponents and doubters of PDP approaches assumes the aspect of a holy war. One reason for this, I suspect, is an implicit adherence to what I shall call the general version of the uniformity assumption. It may be put like this:

> *Every* cognitive achievement is psychologically explicable using only the formal apparatus of a single computational architecture.

I shall say more about the terms of this assumption in due course. The essential idea is easy enough. Some classical cognitivists believe that all cognitive phenomena can be explained by models with a single set of basic types of operation (see the discussion in chapter 1, section 4; in section 5 below; and in chapter 8). This set of basic operations defines a computational architecture in the sense outlined in chapter 1, section 4. Against this view some PDP theorists seem to urge that the kinds of basic operation made available by their models will suffice to construct accurate psychological models of all cognitive phenomena (see chapters 5 and 6 above). Each party to this dispute thus appears to endorse its own version of the uniformity claim. The Classical-cognitivist version is:

> Every cognitive achievement is psychologically explicable by a model that can be described using only the apparatus of classical cognitivism.

The PDP version is:

> Every cognitive achievement is psychologically explicable by a model that can be described using only the apparatus of PDP.

The argument I develop will urge that we resist the uniformity assumption in all its guises. Instead, I endorse a model of mind that consists of a multitude of possibly virtual computational architectures adapted to various task demands. Each task requires psychological models involving distinctive sets of computationally basic operations.

My goal is to cast doubt on the assumption *without* going to the opposite extreme (as is characteristic of the holy-war protagonists) and suggesting that PDP models are *never* psychologically relevant but *always* have the status of mere details of implementation. Both positions are represented in the literature debating the nature of the relation between connectionist models and conventional AI. Thus, Broadbent (1985) argues that psychological explanation involves appeal only to what function is being computed and not to how it is being computed. He further argues that subject to speed constraints, connectionist and conventional AI can both compute all the same functions (i.e., anything that a universal Turing machine can do), so there can be no special psychological interest in developing a connectionist model rather than a more conventional one. Something of this debate has been repeated in a recent exchange in the journal *Cognitive Science* (see Thagard 1986 and Krellenstein 1986). The trouble seems to lie in the ill-defined notion of computing the *same* function. Does a machine that computes "8 × 7" by retrieving a stored answer to "7 × 7" and adding 7 compute the same function as one which adds 7 eight times? They do in the sense that both functions present the same input-output profile (i.e., input: ? 8 × 7, output: 56). But the way they achieve their goal is different. And the difference will affect such fine-grained details as speed and breakdown pattern. The machine that stores the answer to "7 × 7" and adds 7 is in all probability faster. And if through damage it lost its capacity to add, it would still know the answer to "7 × 7" at least, whereas its more conventional cousin would not. Now, it seems reasonable to suppose that one of the tasks of psychological modeling is to offer a computational account of relative-speed profiles, error patterns, and breakdown patterns in human performance. The simple idea that computing the same function is just getting input and output to match for some central range of tasks does not do justice to such fine-grained details. As I remarked earlier in a somewhat different context, it's not what you do, it's how you do it that is of interest, particularly where the how explains a further series of whats. Thus, discrete storage of data and the kind of informationally holistic storage discussed in the previous chapter may well dictate the same range of performance on many tasks. But the speed, pathology, and fine-grained performance (e.g., shading of meanings) of the holistic version may well be different. For this reason one cannot argue from the fact that two systems compute the same function, that PDP models are psychologically irrelevant.

For precisely these kinds of reasons PDP researchers speak of the microstructure of cognition. But this way of speaking is a little dangerous (particularly in the light of their favored analogies). For it begins to suggest a questionable adherence to the uniformity assumption. For example, we saw how, according to their favored analogy, we should conceive the relation between connectionist and conventional AI along the lines of the relation between Newtonian theory and quantum theory.

It is worth repeating a representative passage.

> It might be argued that conventional symbol processing models are macroscopic accounts, analogous to Newtonian mechanics, whereas our models offer more microscopic accounts, analogous to Quantum theory. Note that over much of their range, these two theories make precisely the same predictions about behavior of objects in the world.... However, *in some situations* Newtonian theory breaks down. In these situations we must rely on the microstructural account of Quantum theory. (Rumelhart and McClelland 1986, 125; my emphasis)

This, it seems to me, would be a reasonable analogy just in case in some situations Newtonian theory was held to be the correct physical explanation of the phenomenon in question. But this is not so. Rather, the claim is that in some situations Newtonian theory works, although it is incorrect in detail, and quantum theory explains why. The following uniformity assumption is warranted in the physical case: for every physical event there will exist a (possibly horrendously complex) microstructural quantum-level account that would figure in any *complete* physical model of the phenomenon in question. But what is thus true of the relation between Newtonian theory and quantum theory is not true of the relation between conventional and connectionist accounts as *psychological* models of human achievements, if the main conjecture of this chapter is correct (see the next section).

The claim is often made that "in the subsymbolic paradigm, serial, symbolic descriptions of cognitive processing are approximate descriptions of the higher level properties of connectionist computation" (Smolensky 1987, 103). Or again, "We view macrotheories as approximations to the underlying microstructure which the distributed model ... attempts to capture" (Rumelhart and McClelland 1986, 125). In all these cases my worry is that taken one way, these comments and the general analogy seem to imply that as psychological models, conventional accounts must *always* have the status of (perhaps useful) *approximations* to the real, PDP story. For reasons I develop below, this view threatens to underestimate the value of conventional models and thus invites a holy war that the AI community can ill afford. In the next section I shall argue that the relation between conventional and connectionist AI may not be uniform across the cognitive domain after all. For some tasks, I suggest, the conventional account may be psychologically complete in a way in which Newtonian physics is never physically complete.

3 *Simulating a Von Neumann Architecture*

In a recent paper (Clark 1987) I speculated that the human mind might effectively simulate a serial, symbol-processing, Von Neumann architecture for *some* purposes (largely evolutionarily recent tasks, as discussed in chapter 4). If that is the case, I asked, wouldn't it follow that for *such tasks* some classical-cognitivist computational account could prove to be correct—not just approximately correct but correct, *tout court*? In short, if we suppose that such simulation goes on, isn't the uniformity assumption simply false, and the relation between conventional and connectionist AI not uniform across the cognitive domain?

When writing that paper, I had no idea about how such simulation might occur. I simply relied on the idea that there is nothing unusual in the simulation of one architecture by another. The pervasive idea in computer science of a virtual machine is precisely the idea that a machine can be programmed to behave as if it were operating a different kind of hardware (see, e.g., Tannenbaum 1976). Since then, however, things have come to seem a little more concrete. Thus in Rumelhart, Smolensky, et al. 1986 we find some fascinating speculative ideas on the human capacity to engage in various kinds of conscious, symbolic reasoning. These speculations can be used, I believe, to give substance to the claim that we might occasionally simulate a Von neumann architecture.

The PDP Group raise the following questions. "If the human information-processing system carries out its computations by 'settling' into a solution rather than applying logical operations, why are humans so intelligent? How can we do science, mathematics, logic etc.? How can we do logic if the basic operations are not logical at all?" They suggest an answer that involves a neat computational twist. Our capacity to engage in formal, serial, rule-governed reasoning, they speculate, is a result of "our ability to create artifacts—that is, our ability to create physical representations that we can manipulate in simple ways to get answers to very difficult and abstract problems." (Both passages are from Rumelhart, Smolensky, 1986, 44.)

Before seeing how this solution works, it is worth expanding on the problem for human intelligence and PDP that it is intended to solve. In the first of the two quoted passages the problem is about our ability to do science, mathematics, and logic. Earlier in the same section (Rumelhart, Smolenky, et al. 1986, 38) the problem areas identified are conscious thought, serial processing, and the role of language in thought. Very generally, it seems to me, there are two *kinds* of human capacity that PDP models at first glance are hard put to capture or illuminate. These are:

(1) Processes of serial reasoning in which the ordering of operations is vital,

(2) Processes of generative reasoning in which unbounded set of structures may be produced by the application of rules to a data base.

Conscious planning, logic, and much advanced abstract thought seems to involve capacity 1. The prime example of capacity 2 would seem to be language production. The kind of story discussed below is best suited to explaining the sequential conscious phenomena adverted to in capacity 1. Perhaps it can be extended to cover the generative phenomena as well, but that, I believe, is a much harder issue and one I make no claims to address here. PDP models, like the model of sentence processing discussed in the previous chapter, seem best suited to modeling some aspects of language *understanding*. Language production, insofar as it involves finding and combining the right constituents in the right way to express a message, raises a whole host of other issues that lie largely beyond the scope of this book.

As far as sequential, conscious thought is concerned, PDP approaches so far offer a two-faceted account. First and most simply, there *is* seriality in PDP. Since at one time a network occupies at most one state (a pattern of activation over the simple units), there will be a *sequence* of such states as exogenous and endogenous inputs cause it to become active and then relax into new stable states. Considered during the short periods of value-passing activity, it is a parallel distributed system. Considered over longer stretches of time, it can be seen as a sequence of discrete states. This gives the PDP theorist the beginnings of an angle on conscious experience. The idea being that "the contents of consciousness are dominated by the relatively stable states of the system.... Consciousness consists of a sequence of interpretations—each represented by a stable state of the system" (Rumelhart, Smolensky, McClelland, and Hinton 1986, 39). We are often not conscious of, say, the process of finding a good metaphor, making a pun, various creative leaps in scientific discovery (more on which below). But we are conscious of, say, applying *modus ponens* to two lines of a logical proof or planning a sequence of events. The PDP account would begin to explain this by positing that relaxation occurs during the unconscious fast phenomena and by treating conscious phenomena as a perceived sequence of the results of such relaxation steps.

But the sequentiality of states alone is insufficient to cover even the processes of serial reasoning associated with capacity (1). For, in effect, all we have so far is a kind of stream-of-consciousness display. In cases of logical reasoning, long multiplications and so on, the ordering of operations is vital. How are such orderings achieved? Here is where the second facet of the account comes in, namely, the use of artifacts as physical representations.

Consider a simple example. Suppose I ask you to take every second number in a spoken series, add 2 to it, and sum up the total. Most of us would find the task quite difficult. But suppose I allow you to use pen, paper, and the Arabic numerals. The task becomes simple. For the series 7, 4, 9, 5, 2, 1, 6, 9 isolate every second number (4, 5, 1, 9), add 2 (6, 7, 3, 11), and sum the total (27).

Rumelhart, Smolensky, et al. develop a similar example involving long multiplication. Most of us, they argue, can learn to just see the answer to some basic multiplication questions, e.g., we can just see that 7 × 7 is 49. This, they suggest, is evidence of a pattern-completing mechanism of the usual PDP variety. But for most of us longer multiplications present a different kind of problem. 722 × 942 is hard to do in the head. Instead, we avail ourselves (at least in the first instance—see below) of an external formalism that reduces the bigger task to an iterated series of familiar relaxation steps. We write:

$$\begin{array}{r} 722 \\ \times\, 942 \\ \hline \end{array}$$

and go through a series of simple pattern completing operations (2 × 2, 2 × 2, 2 × 7, etc.), storing the intermediate results on the paper according to a well-devised scheme. In a highly revealing comment, the authors go on to say, "This is *real* symbol processing and, we are beginning to think, the primary symbol processing that we are able to do. Indeed, on this view, the external environment becomes a key extension to our mind" (Rumelhart, Smolensky, et al. 1986, 46). (On the importance of external symbolisms to human thought, I cannot resist quoting a wonderfully self-negating comment made by a student encountering the attractions of PDP for the first time. He said," It was only when I started to write my ideas down that I realized that explicit representations counted for so little at all.") This general strategy of using external representations recalls nicely the earlier argument (chapter 4, section 3) that biologically sound computational accounts need to investigate ways of exploiting environmental structures to aid cognition. Moreover, it suggests an interesting perspective on the Von Neumann architecture itself, more on which shortly.

It is, of course, true that we can learn to do long multiplication in our heads. And here we encounter the final twist to the story. In-the-head sequential reasoning, the authors argue, is made possible only by constructing a *mental model* of the external structures whose actual physical manipulation enabled us to learn to perform such sequential operations in the first place. Thus, they suggest, we may solve syllogisms by constructing Venn diagrams in our heads. We can do this only because we were first able to construct or observe such diagrams in external, physical form.

The notion of a mental model in play here is just the idea of a network that takes as input some specification of an intended action (e.g., multiply

7 × 7 as if using an external medium) and gives as output a resultant state of the imagined world (e.g., an image of the number "49" suitably arranged on paper as part of a longer multiplication). (A more detailed account of how such mental modeling proceeds can be found in the same chapter, p. 40–44.)

In sum, the speculation (and it is no more than a speculation) is that three capacities combine to allow human beings (who are assumed to be PDP devices at root) to perform complex, sequential, symbol-processing tasks. These are:

(1) a basic PDP pattern-matching capacity,

(2) a capacity to mentally model our environment,

(3) a capacity to physically manipulate our real environment, and to perceive the effects of such manipulations (adapted from Rumelhart, Smolensky, et al. 1986, 44).

They add, mysteriously, "especially important is our ability to manipulate the environment so that it comes to represent something," which is enough to raise any philosopher's hackles (see section 4 below).

The thought, anyway, is that capacities 1 to 3 enable us to reduce sequential, symbol-processing problems to a form that is PDP-tractable. On this account the use of the external environment expands both the range of tasks we can conquer and (by enabling us to use relevant mental models) the kinds of mental reasoning of which we are capable. The picture may be captured in a metaphor: Thought parasitises the world and returns, nourished and enlarged, to the head. Rumelhart, Smolensky, et al. seem to assume that this use of the external world always occurs within the experience of an individual human. But there is no reason to rule out the possibility that models of the external would sometimes become incorporated into our innate hardware as a result of the usual processes of natural selection. If so, in some domains we may be born to reason *as if* we had experienced manipulating a real external environment and constructed a mental model of it. This might open the way to some contact with, e.g., Chomskian conjectures concerning language acquisition (although the powerful learning algorithms deployed by PDP seems, if anything, to point in the opposite direction).

Notice, however, that even if some mental models have evolved, our basic architecture would remain a PDP system, though cunningly configured to make possible certain kinds of sequential, symbolic thought. We would not have an architecture purposely built for such reasoning. The historical snowball effect investigated in chapter 4 works to kludge an architecture chosen for speedy perceptual and sensorimotor processing into something capable of some kinds of sequential, conscious reasoning.

All of this suggests an interesting angle on Von Neumann architecture, with its main memory and the logical, manipulative capacities of a CPU. In basic PDP approaches (i.e., ones *not* involving mental modeling) there is, in effect, no distinction between the processing structure and the data being processed. It is the activation pattern of the processing structure that encodes the data. Von Neumann architectures separate the two in just the way an *environmentally embedded* PDP system would separate the external, symbolic structures from the PDP operations on such structures involved in sequential reasoning. Mental modeling is somehow parasitic on this separation. Perhaps, then, it makes sense to see Von Neumann architecture as mistakenly modeling in-the-head computation on computation that in humans consists of both an in-the-head component and (to begin with) an in-the-world component. If so, the mistake of sequential, symbolic cognitivism is to treat *all* thought as depending only on manipulating something like external symbolic structures according to rules. The mistake is to model *all* thought on our gross manipulations of real, external symbolic structures. Thus, we may half agree with Simon's comments: Humans and computer systems "achieve their intelligence by symbolising external and internal situations and events and by manipulating those symbols. They all employ about the same symbol manipulating processes. Perhaps that particular invariance arose because computers were made (unintentionally) in the image of man" (1980, 37). Conventional computers may well have been made in the image of man's exploitation of external, physical structures. The slip is to believe the form of in-the-head operations is entirely given by such a model. The rigidity and limitations of classical cognitivist models and philosophers' reluctance to see such models as intelligent (see chapters 1 to 3) may all stem from this one error. The new start provided by PDP is to see much in-the-head computation as having the qualitatively different form of relaxation procedures and to see other computation as involving the cunning manipulation of external structures (or mental models thereof) to our own ends. In precisely this sense, then, we finally reject the SPSS hypothesis outlined in chapter 1. The manipulation of gross symbolic structures is merely ingenious icing on the computational cake. In the absence of a PDP substrate of powerful pattern-matching operations, such manipulations fail to instantiate thoughts (see section 5 below). We thus uphold Searle's belief that such gross manipulations are not sufficient for thought, though without his mysterious biological alternative (chapter 2).

But for all that, classical cognitivism is not too bad off. For it looks as if the PDP Group's own conjectures undermine the uniformity assumption as a psychological explanation. For a certain range of tasks we can now see the world (or a mental model of it) as the memory of a conventional Von Neumann machine and our physical capacities (in the first instance) as the manipulative capacities of the CPU. In short, for some tasks we seem to simulate a conventional architecture. If so, why shouldn't a correct psychological account of human performance of *such tasks* be given at the classical-cognitivist level? That is to say, why not see the classical account as psychologically accurate in such cases and *not* merely as a good approximation to an accurate account?

An example may help. When we simulate a connectionist architecture on, say, a serial Symbolics LISP machine, we don't speak of the serial account as explaining, in any psychologically relevant way, the microstructure of connectionist cognition. The serial Symbolics machine, in such cases, is indeed just an implementation detail. In the cases in which PDP mind and manipulation of the world yield a virtual Von Neumann architecture, shouldn't we likewise treat the PDP substrate as a psychologically irrelevant implementation detail? If we should, the uniformity assumption fails. For those tasks the correct level of psychological explanation is indeed that of classical cognitivism, though for other tasks it is not.

There are, however, complications. As I remarked earlier, implementation *does* affect the behavior of the simulated machine both in speed and breakdown profiles. I shall return to this point in section 5 below. First, a few worries about the story so far.

4 *A Lacuna in the Account of Real Symbol Processing*

The point of this section is just to emphasize some worries touched on earlier. The path to real-symbol processing seems to have been traversed a little too quickly. Perhaps Rumelhart et al. are correct when they write, "Especially important here is our ability to manipulate the environment so that it comes to represent something." (Rumelhart, Smolensky, et al. 1986, 45). But *how* is this important ability achieved? As far as I can see, they provide no real account.

Consider the example of long multiplication. Most of the work is done by understanding what it is to have an external formalism that represents numbers. Such understanding seems as mysterious in a PDP approach as ever it was in a classical cognitivist one. Monkeys, for instance, seem to have good, basic PDP pattern-matching capacities. And we can give them on a plate, as it were, the external representational formalisms of language and number. But the extent to which they succeed in exploiting such formalisms remains limited. Here is a computational lacuna that will require more than mutterings about mental models to resolve. The question is, What are the prerequisites for a system to come to use external structures in the rich representational fashion of human beings? This question cuts across the boundaries of evolutionary biology, philosophy, and AI, and it is, I suspect, of absolutely the utmost importance. If I had any idea of how to solve it, I'd be writing a very different book.

One thing does seem clear. The apparatus that Rumelhart et al. give won't quite do. The difficulty of seeing how we construct external formalisms in the first place is solved, they say, by the facts of cultural transmission (Rumelhart, Smolensky, et al. 1986, 47). Representational systems, they rightly point out, are not easy to come by. Those we have grew gradually out of a long historical process of alternation and addition to simple systems. This is fine. They are other examples, no doubt, of the pervasive principle of gradualistic holism, examined in chapter 4. But it leaves untouched the real problem of how we can recognize anything as an external representational formalism at all. It is here, I believe, that the most profound problems lie.

And there are further difficulties. Even if one has both an external representational formalism and an understanding of what it is for some squiggles to represent something, still the very deployment of the formalism looks more problematic than Rumelhart et al. allow. Take the example of long multiplication. It is not *simply* a matter of performing an iterated series of basic pattern-matching operations. For as I remarked earlier, we must store the intermediate results on paper (or in our mental model) according to a *well-devised scheme*. That is, if we compute 7 × 7 as part of a long multiplication, we do not simply store 49; rather, we store a "9" in one place in a sequence on paper and carry on 4 to the next operation. What kind of control and storage structures are necessary for this? And does PDP alone enable us to instantiate them?

All these issues, I believe, are important lacunas in speculations about real symbol processing. For the purposes of this chapter, I am assuming they can be overcome. But we should bear them firmly in mind, especially in the light of the suggestion that understanding the mind requires understanding a number of virtual architectures, with correct psychological model associated with each (see below).

5 *Full Simulation, Intuitive Processing, and the Conscious Rule Interpreter*

In the previous chapter (section 4) we met with Smolensky's notion of a virtual machine that he called a conscious rule interpreter. And we saw how, according to Smolensky, classical accounts would be *less approximately* valid for tasks performed using this virtual machine than for tasks run directly on a so-called intuitive processor. Notice, however, that this claim *upholds* the uniformity assumption but weakens its impact by allowing the classical accounts to be accurate models of competence in idealized conditions (again see chapter 6, section 4; for the complete story, see Smolensky 1988).

Smolensky may well be right about the existence of a conscious rule interpreter deployed in many cases of conscious, linguistically or logically formulated reasoning. But if the account in section 3 above is at all correct, this is not yet the whole picture. For it does not allow for the kind of case in which real, external entities (or mental models thereof) play the role of gross symbols manipulated according to explicit, linguistically formulated rules and heuristics. Smolensky thinks that the classical account is approximately valid in certain cases, because within limits the behavior of a PDP system will often match that predicted by a classical model. This match increases (though it is never perfect) as the dimension shift between the entities invoked at the level of task analysis and the semantic interpretation of the units of the network decreases, i.e., the match increases with increasing semantic transparency.

In these cases the approximate validity of conventional models is thus merely a matter of a limited imput-output equivalence. In the special kind of case treated in section 3 above, in contrast, the conventional model is a realistic model of the *processing structure* of a certain extended virtual machine. This is very different from being a model of the input-output structure of a machine. In the special kind of case in which actual discrete environmental structures (or mental models thereof) are manipulated according to explicitly formulated rules or heuristics, we have a virtual machine that recapitulates the processing steps of a conventional model. That is, the *kinds* of operations we perform on real, external symbolic structures (and hence the kinds we use in any mental model of the same) are just the operations found in a conventional processor, e.g., completely copying a symbol from one location to another, deleting, adding whole symbols (e.g., "cup" to a list), and matching whole symbols. In these special cases, therefore, the conventional model is not any kind of *approximation* to the truth; it *is* the truth.

In effect, then, I wish to add a third architectural possibility to a spectrum formulated by Smolensky. Smolensky's picture of the relation of conventional models to PDP models looks like this:

- Tasks involving the intuitive processor are a rough approximation of system performance.
- Tasks involving the conscious rule interpreter are a good approximation of system performance.

To this picture tentatively add:

- Tasks involving the use of external structures as objects for a conscious rule interpreter are an exact description of (possibly virtual) system structure *and* performance.

There are important twists to the story which remain to be described. In particular, we need to consider *mixed tasks*, i.e., tasks (if there are any) that involve more than one of these virtual machines. We also need to consider the qualitative effects of implementing a symbol processor in a parallel distributed architecture. The PDP group's insistence that the cognitivist account, though often useful, must *always* have the status of some more or less accurate approximation to the true computational, psychological story is, I suggest, based on a subtle misreading of the moral of such cases. I return to these matters in the next chapter. For the present, the lesson is straightforward. If the account just developed is at all plausible, any assumptions of uniformity are premature. There may be a full spectrum of relations between PDP and cognitivist models of cognition according to whether the task aspect under study involves intuitive processing, conscious reasoning, or a full simulation of a gross symbol-processing mechanism.

6 *BACON, an Illustration*

Let me illustrate the position outlined above by considering scientific discovery once again. A cognitivist model of scientific discovery is given by the BACON program, outlined in chapter 1, section 4 (for full details see Langley et al. 1987). To recapitulate briefly, BACON derives scientific laws from bodies of data. Roughly, it works on recorded observations of the values of variables and seeks functions relating such values. In its search for such functions it follows some simple heuristics that suggest what functions to try first and how to proceed in cases of difficulty. The program does not start out with any theoretical bias or expectations concerning the outcome; it simply seeks regularities in data. As a kind of control experiment, Simon employed a graduate student who knew nothing of Kepler's third law (see chapter 1) and give him the sets of data upon which Kepler worked.[2] Told to find a function relating one column of figures (in fact, the radii of planetary orbits) to another (in fact, the periods of planetary revolution), it took the student 60 hours to discover the law. The BACON program is much quicker, but the procedure is the same: a serial, heuristic-guided search for a function relating x and y, uninformed by any understanding of the significance of x, y, or the enterprise of scientific investigation itself.

Simon notes, however, that there are elements in the process of real scientific discovery that are not easily amenable to such an approach. He cites the example of Fleming's spotting the significance of the mouldy petri dish. But any classic flash of insight will do. We might point to Stephenson's (allegedly) watching his kettle boil and conceiving the idea of the steam locomotive or someone's studying the behavior of thermodynamic systems and conceiving the ideas behind PDP. These aspects of scientific discovery have two prime characteristics, which should by now attract our attention.

- The flash of insight is typically fast. The idea just comes to us, and we have no conscious experience of working for it.
- The flash of insight involves using rather abstract perceived patterns in one domain of our experience to suggest ways of structuring our ideas about some other apparently far removed domain.

In the light of these characteristics it is not be absurd to suggest that some PDP mechanism is operating in such cases. Simon appears to reject this idea, insisting that these very fast processes are in no way fundamentally different from the processes of serial, heuristic search used by BACON (a full quote is given in chapter 1, section 4). This time, it seems, it may be the conventional theorist who is too quickly assuming uniformity across the cognitive domain.

A better position is well explained by D. Norman: "People interpret the world rapidly, effortlessly. But the development of new ideas, or evaluation of current thoughts proceeds slowly, serially, deliberately. People do seem to have at least two modes of operation, one rapid, efficient, subconscious, the other slow, serial and conscious" (1986, 542). According to this model (which was also endorsed by Smolensky; see chapter 6, section 3), the computational substrate of human thought comprises at least two strands. One, the fast, pattern-seeking operations of a PDP mechanism, the other the slow, serial, gross-symbol-using, heuristic-guided search of classic cognitivism. If my earlier speculations are at all correct, this latter strand may at times be dependent on a virtual symbol-processing architecture, *possibly* created by our capacity to exploit real environmental structures (but the genesis of the simulation is not what is at issue here). Were it not for their strange commitment to a single functional architecture (see chapter 1), such a picture ought to be quite acceptable to theorists like Langley, Simon, Bradshaw, and Zytkow. Certainly, they actually set out to model the processes of slow, conscious, serial reasoning. For example, in defence of the seriality of their approach, they suggest that "for the main processes *in any task requiring conscious attention*, short term memory serves as a severe bottleneck that forces processes to be executed serially" (Langley et al. 1987, 113; my emphasis). It seems, then, as if the folk-psychological category of "scientific discovery" fails to pick out a single computational kind. Instead, it papers over the difference between conscious toil and unconscious insight to fix on the *product* of both processes (new scientific ideas), a product of undoubted significance in human life. This is just what the conjectures of chapter 3 should lead us to expect. The folk-psychological view of the cognitive terrain is a view from within the environmentally and culturally rich mesh of human practices. It has no special interest in fixing on regularities and differences in the causal computational substrate of our thought except insofar as these are of immediate practical significance.

Suppose, then, that we were to accept such a division within the domain of scientific discovery. Would it not then be fair to seek psychological models of the slow, serial component within a classic cognitivist framework and models of the fast component within PDP? An *overall* model of human scientific thinking needs to include both and to address the issue of how the results of each can be fed to the other in a cooperative way. Still, it would not follow that the slow serial model is a mere approximation of that component.

In sum, I am advocating that cognitive science is an investigation of a mind composed of many interrelating virtual machines with correct psychological models at each level and further accounts required for the interrelations between such levels. Only recognition of this multiplicty of mind, I suspect, will save cognitive science from a costly holy war between the proponents of PDP and the advocates of more conventional approaches.[3]

Chapter 8
Structured Thought, Part 1

1 *Weighting for Godot?*

Are some cognitive competences beyond the explanatory reach of *any* PDP model? Does the weighting game degenerate at some specifiable point into waiting for Godot?[1] Some philosophers and cognitive scientists believe so. The putative problem concerns the systematicity of the processing required for such sophisticated cognitive achievements as language production and understanding. Two kinds of argument are advanced to convince us that connectionism is unable to penetrate these systematic domains. One is a lively (but ultimately implausible) set of arguments detailed in Fodor and Pylyshyn 1988 and Fodor 1987. These arguments seek to support a classical-cognitivist model of thought. The other, associated with an influential critique of connectionism by Pinker and Prince (1988), aims to justify at least a classical structuring of the components of information-processing models of higher cognitive functions. The arguments here are plausible and important but, I shall argue, are unable to support any strong conclusions about the limits of PDP.

The itinerary goes like this. The current chapter focuses on the arguments of Fodor and Pylyshyn. These are shown to be generally uncompelling. Systematicity of effect may well argue in favor of systematicity of cause. But classical cognitivism involves much larger claims, which Fodor and Pylyshyn give us no reason to accept. The chapter ends by associating their error with a pervasive failure within the cognitive-science community to distinguish two *kinds* of cognitive science. One of these involves the attempt to model the complex, holistic structure of thought (thoughts here simply *are* the contentful states described using propositional-attitude ascriptions). The other is the attempt to develop models of the in-the-head, computational causes of the intelligent behavior that *warrants* such thought ascriptions. The projects, I shall argue, are distinct and nonisomorphic. Chapter 9 goes on to consider some edited highlights of the Pinker and Prince paper and then returns afresh to the question of mixed models of cognitive processes, first raised in chapter 7.

2 *The Systematicity Argument*

Fodor and Pylyshyn 1988 is a powerful and provocative critique aimed at the very foundations of the connectionist program. In effect, they offer the friend of connectionism an apparently fatal dilemma. Either connectionism constitutes a distinctive but inadequate cognitive model, or if it constitutes an adequate cognitive model, it must do so by specifying an implementation of distinctively *classical* processing strategies and data structures. I shall argue that the critique of Fodor and Pylyshyn is based on a deep philosophical confusion.

I begin with an imaginary anecdote, the point of which should become apparent in due course. One day, a famous group of AI workers announced the unveiling of the world's first, genuine, thinking robot. This robot, it was claimed, *really* had beliefs. The great day arrived when the robot was put on public trial. But there was disappointment. All the robot could do, it seemed, was output a single sentence: "The cat is on the mat." (It was certainly a sophisticated machine, since it generally responded with the sentence when and only when it was in the presence of a cat on a mat.) Here is an extract from a subsequent interchange between the designers of the robot and some influential members of the mildly outraged academic community.

> *Designers:* Perhaps we exaggerated a little. But it really *is* a thinking robot. It really does have at least the *single* belief that the cat is on the mat.
>
> *Scoffers:* How can you say that? Imagine if you had a child and it could produce the sentence "The cat is on the mat" but could not use the words "cat," "on," and "mat" in any other ways. Surely, you would conclude that the child had not yet learned the meaning of the words involved.
>
> *Designers:* Yes, but the child could still *think* that the cat is on the mat, even if she has not yet learned the meanings of the words.
>
> *Scoffers:* Agreed, but the case of your robot is even worse. The child would at least be capable of appropriate perceptual and behavioral responses to other situations like the mat being on the cat. Your robot exhibits no such responses.
>
> *Designers:* Now you are just being a behaviorist. We thought all that stuff was discredited years ago. Our robot, we can assure you, has a data structure in its memory, and that structure consists of a set of distinct physical tokens. One token stands for "the," one for "cat," one for "is," one for "on," and one for "mat." Unless you're behaviorists, why ask more of a thought than that?
>
> *Scoffers:* Behaviorists or not, we can't agree. To us, it is *constitutive* of

> having the thought that *a* is *b* to be able to have other thoughts involving *a* and *b*, e.g., *c* is *b*, *a* is *d*, *a* is not *b*, and so on. To have a thought is to be in a state properly described by the ascription of a set of concepts and relations. And you can't have a concept in a *semantic vacuum*. You can't know what addition is if *all* you can do is output "2 + 2 = 4." Possession of a concept involves a large and structured set of abilities to *do* things, either internally or externally. This is not any kind of peripheral behaviorism; it's just a reflection of the actual nature of thought ascription. As a matter of fact, we think that's what the business of thought ascription is all about. It's a way of making global sense of a *whole set* of dispositions to behave.

The moral of the story is just this. You can't get away with ascribing the thought that *a* is *b* to a system unless you can also get away with ascribing *other* thoughts involving *a* and *b* to it (in actual *or* counterfactual circumstances). (This observation pervades much recent philosophy of language. The general, global, holistic nature of belief ascription is well described in various works by D. Davidson (see the Davidson (1984) collection). And the point about the need to be able to entertain *many* thoughts involving *a* and *b* to be capable of entertaining any is made explicit as the generality constraint in Evans 1982, 100–105 (see chapter three above).

Consider now the mainstay of Fodor and Pylyshyn's assault on connectionism: the requirement of systematicity. The argument goes like this.

> *Observation:* Normal linguistic competence of a native speaker is *systematic*. Speakers who know how to say "John loves the girl" generally also know how to say "The girl loves John."
>
> *Explanation:* Linguistic competence involves grasp of compositional semantics. The speaker learns to construct meaningful sentences by combining meaningful atomic parts in a particular way. Thus, competence with "John," "loves," "the," and "girl," along with competence with subject-verb-object constructions, immediately yields a capacity to produce "The girl loves John."

Sentences have genuine constituent and constructive structure, and this fact *explains* the phenomenon of systematicity.

Fodor and Pylyshyn then propose an exactly analogous argument for *thought*. It goes like this.

> *Observation:* Normal (human and animal) cognitive competence is systematic. You don't find creatures with punctate minds, e.g., creatures whose cognitive capacities "consist of the ability to think seventy-four unrelated thoughts" (Fodor and Pylyshyn 1988, 40). Creatures who can *think* that John loves the girl can typically also think that the girl loves John.

> *Explanation:* Thoughts, like sentences, have constituent structure. Thinking that John loves the girl involves having some relation to an internal representational structure with proper parts standing for "John," "loves," "the," and "girl" and with some kind of combinatorial syntactic structuring.

In sum, there will be two mental representations, one corresponding to the thought that John loves the girl and one corresponding to the thought that the girl loves John. And there will be some systematic relation between them such that "the two mental representations, like the two sentences, must be made of the same parts' (Fodor and Pylyshyn 1988, 39). Fodor and Pylyshyn conclude: "If this explanation is right (and there don't seem to be any others on offer), then mental representations have internal structure and there is a language of thought. So the architecture of the mind is not a connectionist network" (1988, 40). The anticonnectionist conclusion is not yet compelling, of course. It depends on a lemma to the effect that connectionist work, in contrast, posits *unstructured* mental representations. Fodor and Pylyshyn certainly endorse such a lemma. They write, "Connectionists propose to design systems that can exhibit intelligent behavior without storing, retrieving or otherwise operating on structured symbolic expressions (1988, 5). (Call this the lemma of unstructured representations.)

The overall form of argument is now visible.

> Thought is systematic.
>
> So internal representations are structured.
>
> Connectionist models posit unstructured representations.
>
> So connectionist accounts are inadequate as distinctive cognitive models.

Classical accounts, by contrast, are said to posit internal representations with rich syntactic and semantic structure. They thus reach the cognitive parts that connectionists cannot reach.

3 *Systematicity and Structured Behavior*

This argument is deeply flawed in at least two places. First, it misconceives the nature of thought ascription, and with it the significance of systematicity. Second, it mistakenly infers lack of compositional, generative structure from lack of what I shall call conceptual-level compositional structure. These two mistakes turn out to be quite interestingly related.

The point to notice on the nature of thought ascription is that systematicity, as far as Fodor and Pylyshyn are concerned, is a *contingent, empirical* fact. This is quite clear from their discussion of the systematicity of infraverbal, animal thought. Animals that can think *aRb*, they claim, can generally think *bRa* also. But they allow that this *need not* be so. "It is," they write, "an empirical question whether the cognitive capacities of infraverbal organisms are often structured that way" (1988, 41). Now, it is certainly true that an animal might be able to respond to *aRb* and not to *bRa*. But my claim is that in such a case (*ceteris paribus*) we should conclude *not* that it has, say, the thought "*a* is taller than *b*" but cannot have the thought "*b* is taller than *a*." Rather, its patent incapacity to have a *spectrum* of thoughts involving *a*, *b*, and the taller-than relation should defeat the attempt to ascribe to it the thought that *a* is taller than *b* in the first place. Perhaps it has a thought we might try to describe as the thought that *a*-is-taller-than-*b*. But it does *not* have the thought reported with the ordinary sentential apparatus of our language. For grasp of such a thought requires a grasp of its component concepts, and that requires satisfying the generality constraint.

In short, Fodor and Pylyshyn's "empirical" observation that you don't find creatures whose mental life consists of seventy-four unrelated thoughts is no *empirical* fact at all. It is a *conceptual* fact, just as the "thinking" robot's failure to have a single, isolated thought is a conceptual fact. Indeed, the one is just a limiting case of the other. A radically punctate mind is no mind at all.

These observations should begin to give us a handle on the *actual* nature of thought ascription. Thought ascription, as we saw in chapter 3, is a means of making sense of a *whole body* of behavior (actual and counterfactual). We ascribe a *network* of thoughts to account for and describe a rich variety of behavioral responses. This picture of thought ascription echoes the claims made in Dennett (1981). The folk-psychological practice of thought ascription, he suggests, "might best be viewed as a rationalistic calculus of interpretation and prediction—an idealizing, abstract, instrumentalistic interpretation method that has evolved because it works" (1981, 48). If we put aside the irrealistic overtones of the term "instrumentalism" (a move Dennett himself now approves of—see Dennett 1987, 69–81), the general idea is that thought ascription is an "abstract, idealising, holistic" process, which therefore need not correspond in any simple way to the details of any story of in-the-head processing. The latter story is to be told by what Dennett (1981) calls "sub-personal cognitive psychology." In short, there need be no neat and tidy quasireductive biconditional linking in-the-head processing to the sentential ascriptions of belief and thought made in daily language. Instead, a subtle story about in-the-head processing must explain a rich body of behavior (actual and counterfactual, external and internal), which we *then* make holistic sense of by ascribing a systematic network of abstract thoughts.

It may now seem that we have succeeded in merely relocating the systematicity that Fodor and Pylyshyn require. For though it is a conceptual fact, and hence as unmysterious to a connectionist as to a classicist, that *thoughts* are systematic, it is a plain old *empircal* fact that behavior (which holistically warrants thought ascriptions) is generally as systematic as it is. If behavior wasn't systematic, the upshot would be, not punctate minds, but no minds. But that it *is* systematic is an empirical fact in need of explanation. That explanation, according to Fodor and Pylyshyn, will involve wheeling out the symbolic combinatorial apparatus of classical AI. So doesn't the classicist win, though one level down, so to speak? No, at least not without an *independent* argument for what I called conceptual-level compositional *structure*. It's time to say what that means.

One pivotal difference between classical accounts and those that are genuinely and distinctively connectionist lies, according to Fodor and Pylyshyn, in the nature of the internal representations they posit. Recall that classicists posit internal representations that have a semantic and syntactic structure similar to the sentences of a natural language. This is often put as the claim that "classical theories—but not connectionist theories—postulate a 'language of thought'" (Fodor and Pylyshyn 1988, 12). And what that amounts to is at least that the internal representation, like a sentence of natural language, be composed of parts that, together with syntactic rules, determine the meanings of the complex strings in which they figure. It is further presumed that these parts will more or less line up with the very words that figure in the sentences that report the thoughts. Thus, to have the thought that John loves the girl is to stand in some relation to a complex internal token whose proper parts have the context-independent meanings of "John," "loves," and so on. This is what it is to have a conceptual-level compositional semantics for internal representations. Distributed connectionists, in contrast, were seen not to posit recurrent internal items that line up with the parts of conceptual-level descriptions. Thus, "The coffee is in the cup" would, we saw, have a subpattern that stands for "coffee." But that subpattern will be heavily dependent on context and will involve microfeatures that are specific to the in-the-cup context. We need not dwell further on the details of this difference here. For our purposes, the important point is simply this: There is no independent argument for the conceptual-level compositionality of internal representations. And without one, systematicity does not count against connectionism.

Let us see how this point works. Fodor and Pylyshyn require a kind of systematicity that argues for a language of thought, i.e., for a system of internal representations with conceptual-level compositionality. One approximation to such an argument in their text is the following: "It is ... only insofar as 'the' 'girl' 'loves' and 'John' make the same semantic contribution to 'John loves the girl' that they make to 'The girl loves John' that understanding the one sentence implies understanding the other" (Fodor and Pylyshyn 1988, 42). If the locus of systematicity in need of explanation lay in thought-ascribing *sentences,* then this would indeed constitute the required argument. But the systematicity of thought-ascribing sentences is, we saw, a *conceptual* matter. Finding thoughts there at all requires that the ascriptive sentences form a highly structured network. What is *not* a conceptual matter is the systematicity of the *behavior* that holistically *warrants* ascriptions of thoughts. But here there is no obvious pressure for a system of internal representations that *themselves* have conceptual-level systematicity. All we need is to be shown an internal organization that explains why a being able interestingly to respond to a blue square inside a yellow triangle, for example, should also be able interestingly to respond to a yellow square in a blue triangle. And connectionist models, invoking, e.g., various geometric microfeatures as a means of identifying squares and triangles, can do just this. And they can do so even if the resultant system has no single internal state that constitutes a recurrent and context-independent representation of "square" "triangle," and so on. Likewise, in the room example (reported in chapter 5, section 4) we are shown a model that can represent bedrooms and living rooms as sets of microfeatures. And it is no mysterious coincidence that the model could thereby represent a large fancy bedroom (one with a sofa in it). It could do so because of the recurrence of many microfeatures across all three cases. Highly distributed, microfeatural systems will thus exhibit all kinds of systematic behavioral competence *without* that competence requiring explanation in terms of conceptual-level compositionality.

In sum, the systematicity of thoughts is a conceptual requirement if we are to be justified in finding thoughts there at all. What stands in need of empirical explanation is not the systematicity of thoughts but the systematicity of the behavior, which grounds thought ascription. Such systematicity indeed suggests recurrent and recombinable elements. But there is no reason to suppose these have to have a conceptual-level semantics. (Indeed, given the holistic nature of the thought ascriptions from which conceptual-level entities are drawn, this looks unlikely.)

The lemma of unstructured representations, upon which the anticonnectionist force of the argument from systematicity depends, is thus unsupported. All that is supported is, if you like, a lemma of no conceptual-level structure. But once we cease to be blinded by the glare of sentential thought ascriptions, any lack of conceptual-level structure ceases to be a problem and begins to look suspiciously like an advantage.

This shows, incidentally, why a certain kind of defense of Fodor and Pylyshyn's position won't work. The defense (put to me by Ned Block) claims that the implausibility of finding neat in-the-head correlates to conceptual-level structures is irrelevent to the claims of a language of thought. For what the language-of-thought hypothesis claims, according to this defense, is just that there must be *some* systematic description of the thought that figures in a computational account of the mind of the thinker. In other words, if there is a systematic description in English of a creature's thought, we need to postulate an internal language that corresponds point by point to *some* systematic description of that thought, though it need not correspond to the particular description given by the English sentence.

Such a defence won't work against the criticism I am advancing. For my criticism flows directly from the picture of thought ascription developed above (and in chapter 3). The upshot of this position is that as a holist (and indeed, a kind of behaviorist) about thought ascription, I would deny that *any* systematic description of a thought as picked out by our daily talk of thoughts is likely to be a good guide to actual in-the-head processing. This is because my account drives a wedge between real in-the-head states and thought reports. The items that our daily talk picks out as thoughts are not, on my account, good candidates for instantaneous brain states. So even if you redescribe those items in some other way, the problem remains. For it is a problem about the very objects in need of computational explanation. I deny that daily thought reports isolate those objects. As a result, my scepticism stands even against Block's version of the systematicity claim. *No* description of thoughts reported in our daily talk will capture systematic facts about in-the-head processing.

4 *Cognitive Architecture*

Fodor and Pylyshyn also criticize connectionists for confusing the level of psychological explanation and the level of implementation. Of course, the brain is a connectionist machine at one level they say. But that level may not be identical with the level of description that should occupy anyone interested in our *cognitive architecture*. For the latter may be best described in the terms appropriate to some virtual machine (a classical one, they believe) implemented on a connectionist substructure. A cognitive architecture, we are told, "consists of the set of basic operations, resources, functions, principles, etc. ... whose domain and range are the *representational states* of the organism" (Fodor and Pylyshyn, 1988, 10). Fodor and Pylyshyn's claim is that such operations, resources, etc. are fundamentally classical; they consist of structure-sensitive processes defined over internal, classical, conceptual-level representations. Thus, if we were convinced of the need for classical representations and processes, the mere fact that the brain is a kind of connectionist network ought not to impress us. Connectionist architectures can be implemented in classical machines and vice versa.

This argument in its pure form need not concern us if we totally reject Fodor and Pylyshyn's reasons for believing in classical representations and processes. But it is, I think, worth pausing to note that an intermediate position is possible. Suppose we accepted the idea (suggested in chapter 7) that for some purposes at least the brain simulates a classical machine using a connectionist substructure. Even then, I suggest, it doesn't follow that the connectionist substructure constitutes psychologically irrelevant implementation detail. For one benefit of connectionist research has surely been to show how psychologically interesting properties can emerge out of what looks like mere implementation detail from a classical perspective.

To take an example, consider Smolensky's (1988, 13) idea of a "subconceptually implemented rule-interpreter." This is, in effect, a classical symbol processor implemented using a connectionist machine. Now consider the task of generating mathematical proofs. Implementing the classical rule interpreter on a larger connectionist substructure, Smolensky suggests, might permit it to access and use some characteristic PDP operations. For example, it may generate the applicable rule by using a flexible, context-sensitive, best-match procedure. Once generated, however, the rules could be applied rigidly and in serial fashion by the classical virtual machine. Thus, "the serial search through the space of possible steps that is necessary in a purely symbolic approach is replaced by intuitive generation of possibilities. Yet the precise adherence to strict inference rules that is demanded by the task can be enforced by the rule interpreter; the creativity of intuition can be exploited while its unreliability can be controlled" (Smolensky 1988, 13).

It seems, then, that a connectionist implementation of a classical machine (the rule interpreter) may involve representations in ways that, if they are to be explained at all, require reference to the details of connectionist modes of storage and recall. Fodor and Pylyshyn can of course, deny that the explanation of such properties is a matter of proper psychological interest. But this surely does not ring true. To take just a single case, to explain semantically specific deficits (roughly, aphasias in which *classes* of knowledge are differentially imparied—e.g., the loss of names of indoor objects or of fruits and vegetables), Warrington and McCarthy (1987) propose an explanation that is both connectionist (or associationist) in spirit and depends on factors that, to a classicist, would look remarkably like "mere implementation detail." Roughly, they suggest that a connectionist model of the *development* and *storage* of semantic knowledge may account for the fractionations observed.

At the very least, it is surely a mistake to think (as Fodor and Pylyshyn appear to do) in terms of one task, one cognitive model. For (as suggested in chapter 7) our performance of any top-level task (e.g., mathematical proof) may require computational explanation in terms of a number of possibly interacting virtual machines, some classical, some connectionist, and some of an unknown architecture. This multiplicity will be mirrored in the explanation of various task-related pathologies, aphasias, and so on.

What is implementation detail relative to one aspect of our performance of a particular task may be highly relevant (psychological, representation-involving) detail relative to other aspects of the same task.

What this opens up is the possibility of a partial reconciliation between the proponents of classical AI and connectionists. For some aspects of our performance of some tasks, it may well be entirely correct (i.e., nonapproximate) and necessary to couch a psychological explanation in classical terms. Perhaps some of Fodor and Pylyshyn's arguments serve to pick out those aspects of human performance that involve the use of a virtual classical machine (for example, their comments on the difficulties that beset certain kinds of logical inference if we use context-relative representations [1988, 46]). But since their completely general argument for the systematicity of thought fails, these cases all constitute a much smaller part of human cognition than they expect. And even when a classical virtual machine is somehow implicated in our processing, its operation may be deeply and inextricably interwoven with the operation of various connectionist machines.

In a recent talk, G. Hinton hinted at a picture of high-level cognition that would facilitate just such a reconciliation. The idea is to give a system two different internal representations of everything. One of these would be a single feature representing, e.g., Mary (Hinton calls this the reduced description). The other would be a fully articulated microfeatural representation of Mary (the so-called expanded description). And it would be possible to go on from one of these to the other in some nonarbitrary fashion. The idea was not elaborated, and I shall not speculate too much on it here. But it is worth just noting that such a model would seem to provide (in the reduced descriptions) the kind of data structure upon which any virtual classical machine would need to operate. But it provides such structures *within* the context of an overall system in which the existence and availability of the expanded, connectionist representation is presumably crucial to many aspects of performance.

5 *Two Kinds of Cognitive Science*

Fodor and Pylyshyn's mistake is to project the systematicity of ascriptions of propositional attitudes directly back onto a matching syntactic systematicity in brain computation. The root cause of this is a failure to understand the nature and goals of thought talk itself. (Again, recall chapter 3.) This same mistake is, I believe, the cause of an upsetting pathology within the cognitive science community. The pathology makes itself felt in all-too-frequent exchanges along the following lines: in discussion following a paper on an AI model of some aspect of human thought a questioner addresses the speaker.

Questioner: Are you claiming that that's how human beings think then?

Speaker: Oh no. Absolutely not. *Of course,* our brains don't use a predicate calculus. In fact, it's unlikely that the algorithms we use bear any relation to logical calculi at all.

Questioner: So your project is really part of technological AI. You want some program to get a certain input-output mapping right, but you don't really care about how humans think.

Speaker: Well, not exactly. Really, it's hard to say *what* the project is, because it *is* human thought that we're trying to model.

Such exchanges are by no means uncommon. They are not restricted to fledgling AI workers, nor are they the exclusive hallmark of conventional cognitivists.

Here is an admittedly highly speculative hypothesis about the cause of such confusion. The received wisdom is that AI comes in two varieties: *technological* AI, in which the goal is simply to get a machine to do something with no commitment to producing a model of human psychology, and *psychological* AI (or cognitive science), in which the goal is to produce a computational model of human or animal psychological states and processes.

But suppose that the arguments developed in chapter 3 have some force. Suppose, that is, that thought ascription is essentially a matter of imposing a holistic interpretation upon a large body of behavior in an environmental context. The individual thoughts thus ascribed are perfectly real, but they are not the kind of entities that have neat, projectible, computational analogue in the brain. What then becomes of the project of psychological AI or, more generally, cognitive science?

The radical conjecture I would like briefly to pursue is this. Cognitive science turns out to encompass *two* projects, each laudable and legitimate but absolutely distinct. These projects coincide with two ways of understanding the notion of a psychological model. A psychological model may be a model of the complex structure of human (or animal) thought i.e., the holistic network of ascriptions of contentful states. Or it could be a model of the computational operations in the brain that in part make possible the rich and varied behavior we describe using propositional-attitude talk. Contrary to what Fodor thinks, these models will typically be nonisomorphic, though there may be exceptions (see below). In short, there are two kinds of cognitive science: descriptive cognitive science and causal cognitive science.

Descriptive cognitive science attempts to give a formal theory or model of the structure of the abstract domain of thoughts, using the computer program as a tool or medium.

> *Causal cognitive science* attempts to give an account of the inner computational causes of the intelligent behaviors that form the basis for the ascription of thoughts.

Notice that descriptive cognitive science is *not* the same as technological AI, since descriptive cognitive science cares a great deal about actual human psychology. Nor is it the same as what Searle (1980) calls "weak AI." For weak AI uses the computer as a tool to develop formal models of brain causes of intelligent behavior. Weak AI is a weak version of causal cognitive science that does not believe that merely instantiating a formal model of the brain causes of intelligent behavior will yield a system with intentional states (see chapter 2). Incidentally, the distinction between descriptive and causal cognitive science yields quite a neat perspective on Searle's general orientation. For Searle objects that a computer manipulating formal tokens is at most manipulating the formal shadows of thought (again, see chapter 2). We can now reconstrue this claim as a claim that manipulating the descriptions of thought contents is very different from replicating the inner causes of intelligent behavior. Running a program written according to the paradigm of descriptive cognitive science is *not* the way to instantiate thoughts. Searle is right to see this but wrong to conclude that the difference lies in our biological makeup.

My radical claim, then, is that a great deal of good and important work within cognitive science is *unwittingly* descriptive cognitive science. Failure to distinguish the two projects leads to the confusion and disarray depicted in my imagined dialogue. Anyone engaged in the descriptive task *is* primarily interested in human thought but nonetheless *isn't* giving a model of the computational brain causes of intelligent behavior. To make this idea of a descriptive cognitive science clearer, I shall first draw a parallel with the study of grammar and then offer some examples of the approach in action.

6 Grammars, Rules, and Descriptivism

The cognitive status of grammar is a vexed question to which I do not pretend to do justice here. Instead, my purpose is to use *existing* positions on the status of grammar as an illustration of descriptive cognitive science.

There are three broad positions on the nature of a grammar for a natural language. The first is propositional psychological realism.

> If *a* is a competent speaker of a language, *a*'s competence is causally explained by unconscious knowledge of the rules of a grammar for the language. These rules are internally represented by structures in *a*'s head that have the syntax of the natural language sentences describing the rules.

Propositional psychological realism thus says that a good grammar should be psychologically real and written in the form of explicit rules in a language of thought.

The second position is structural psychological realism.

> If *a* is a competent speaker of a language, *a*'s competence is causally explained by the fact that *a*'s information-processing capacities are structured in a way suggested by the form of a grammar for the language.

Structural psychological realism drops the requirement of explicit, sentential coding of grammatical rules. It merely requires that a good grammar fix (perhaps nonuniquely) on a functional decomposition of the language production system of the brain. For example, if the grammar involves a system of rules for regular cases and a list of exceptions to such rules, it will be good just in case there is a true in-the-head, information-processing story that posits both a system of lexical access that lists exceptions and a distinct, nonlexical component. Katz adopted this position when he wrote: "Componential distinctions between ... syntactic, phonological and semantic components must rest on relevant differences between three neural submechanisms of the mechanism which stores the linguistic description. The rules of each component must have their psychological reality in the input-output operations of the computing machinery of this mechanism" (1964, 133).

The third position on the nature of grammar for natural language is

> A good grammar for a language is any theory that yields all and only the sentences characterized as grammatical by a competent speaker of the language.[4] Such a grammar need not be unique, nor need it suggest the form or content of any psychologically realistic theory of language production or understanding.

(Note that "psychologically realistic" here and elsewhere means "fixing on actual structural or computational features of in-the-head processing." Because of the ambiguity of the notion of a psychological model, this usage could be misleading. But it is standard in the literature, and it seems to be undue nit-picking to demur.) Descriptivism has no truck with the questions of psychological reality at all. It treats the set of grammatical sentences as given and seeks a formal characterization of that set. A robot appraised of such a characterization would produce only grammatical utterances. But human language production and understanding could work along entirely different lines, and the descriptivist would be unphased. Such a position is adopted by Stich (1972) and more recently by Devitt and Sterelny (1987). Thus, Stich writes that a grammar "describes certain language-specific facts: facts about the acceptability of expressions to speakers and facts about an ability or capacity speakers have for judging and classifying expressions as having or lacking grammatical properties and relations.... It is perhaps misleading to describe [the grammarian] as constructing a theory of the language of his subjects. Rather he is building a description of the facts of acceptability and linguistic intuition" (1972, 219–220).

The debate between the psychological realists and the descriptivists has on occasion been needlessly acrimonious. The realists (Fodor [1980b], Chomsky and Katz [1974]), accuse the descriptivists of flouting general canons of scientific practice. The descriptivists (Stich [1971, 1972], Devitt and Sterelny [1987, 142–146]) accuse the realists of misapplying inference to the best explanation. The debate (nicely summarized in Devitt and Sterelny 1987) has gone roughly like this. First volley: "A good grammar for English will generate all and only the grammatical strings of English. The best explanation of the fact that competent speakers generate and judge grammatical just grammatical strings is that they internally represent the grammar in some way. Grammar is thus psychologically real." The descriptivists return the ball by saying, "Psycholinguistic evidence seems to play little role, so far, in the determination of grammars. If that pattern continues, a good grammar may well be found that is simple, elegant, and gets the right strings. But why treat this as a basis for an inference to the psychological reality of grammar? After all,

> First, we would want evidence that G [the grammar] was a *candidate* for psychological implementation; that the transformational processes it implicated were within the computational ambit of the mind. Second, the very elegance and simplicity of G is rather more evidence *against*, than evidence for, it being the grammar our brain is built to use ..., [since] adaptations are typically *not* maximally efficient engineering solutions to the problems they solve. Finally, ... the fact that G is maximally efficient and elegant from the grammarian's point of view does not entitle us to suppose it is optimal from the brain's point of view. (Devitt and Sterelny 1987, 145–146)

Notice how the descriptivists' response fits with the general principles of evolutionary design raised in chapter 4. And notice also that the final point concerning the need to gear psychologically realistic models to constraints imposed by the structure of the brain is highly conducive to a connectionist approach to modeling the brain basis of grammatical competence.

For all that, however, the dispute ought not to call for a choice. Rather, we should conclude only the following: (1) The grammars actually being constructed by working linguists are unlikely to be psychologically real. Nonetheless, they are useful descriptions of real properties of natural languages. (2) Theorists whose goal is the construction of models of the brain basis of grammatical competence will need to focus not only on the data and grammars of (1) but also on the structure of the brain, psycholinguistic evidence, and even, perhaps, evolutionary conjectures concerning the origins of speech and language (see, e.g., Tennant 1984). In short, what is needed is clarity concerning the *goals* of various studies, not a victory of one choice of study over another. Devitt and Sterelny strike a nice balance, concluding that linguists are usefully studying not internal mechanisms but "the truth-conditionally relevant syntactic properties of linguistic symbols" (1984, 146), while nonetheless allowing that such studies may illuminate some general features of internal mechanisms and hence (quite apart from their intrinsic interest) may still be of *use* to the theorist concerned with brain structures.

What is thus true of the study of grammar is equally true, I suggest, of the study of thought. Contentful thought is what is described by propositional-attitude ascriptions. These ascriptions constitute a class of objects susceptible to various formal treatments, just as the sentences judged grammatical constitute a class of objects susceptible to various formal treatments. In both cases, computational approaches can help suggest and test such treatments. But in both cases these computational treatments and a psychologically realistic story about the brain basis of sentence production or holding propositional attitudes may be expected to come apart.

7 *Is Naive Physics in the Head?*

There is a type of work within cognitive science known variously as naive physics, qualitative reasoning, or the formalization of commonsense knowledge. I want to end this chapter by suggesting that most of this work, at least in its classical cognitivist incarnations, may fall under the umbrella of what I am calling descriptive cognitive science. If so, this is a clear case in which descriptive cognitive science and causal cognitive science have got badly confused. For people working in the field of naive physics typically conceive of their work as very psychologically realistic, in contradistinction to much other AI work.

The basic idea behind naive physics is simple and has already been mentioned in chapter 3, section 6. Naive physics is an attempt to capture the kind of commonsense knowledge that mobile, embodied beings need to get around in the real world. We all know a lot about tension and rigidity: you can't push an object with a piece of string. And we know about liquidity, solidity, elasticity, spreading, and so on. The list is endless. And the knowledge is essential. Without it we couldn't spread marmite on toast, predict that beer will spill off a smooth unbounded table top, or drag a shopping trolley along a bumpy road. But the project is still underspecified. I gave the goal as that of capturing commonsense knowledge. But the metaphor of capturing is always dangerous, for it leaves the criteria of success deeply obscure. In the present case "capturing commonsense knowledge" could mean either exhibiting the structure of the set of knowledge ascriptions warranted by a being's practical capacities to get around in the world (the descriptive option), or exhibiting the structure or program of a computational brain mechanism that enables the being to get around the world (the causal option).

There can be little doubt about which project most naive physicists take themselves to be engaged in. Here are a few quotes from recent articles. "We want the overall pattern of consequences produced by [our] theory to correspond reasonably faithfully to our own intuition in both breadth and detail. *Given the hypothesis that our own intuition is itself realised as a theory of this kind inside our heads,* the [naive-physics] theory we construct will be equipotent with this inner theory" (Hayes 1985a, 5; my emphasis). "We should ... concentrate ... on the details of what must be in the heads of thinkers.... When we know what it is that people know, we can begin to make realistic theories about how they work. *Because they work largely by using this knowledge*" (Hayes 1985a, 35; my emphasis). "The motivations for developing a qualitative physics stem from outstanding problems in psychology, education, artificial intelligence and physics. *We want to identify the core knowledge* that underlies physical intuition" (de Kleer and Brown 1985, 109; my emphasis).

Examples could be multiplied (see, e.g., articles in Hobbs and Moore 1985 and in Hallam and Mellish 1987). Not *all* naive physicists insist on psychological reality as much as Hayes. J. R. Hobbs writes, "We, at least in the short term, are happy to have any theory, regardless of how accurately it models people, provided it is formally adequate" (1985, p. xvi). The same slant is evident in David Israel's well-motivated question, "Psychologically realistic or not, are logical formalisms appropriate media for representing our commonsense knowledge of the world?" (1985, 430). Both Hobbs and Israel are thus open to viewing their enterprise as descriptive cognitive science.

That said, many naive physicists do take themselves to be modeling computational brain processes. Yet there is surely room here for some doubts precisely analogous to those raised in the previous section in the context of grammatical competence. Just as in the case of grammar, we face a situation in which human agents are visibly competent at a certain kind of problem solving. And just as we can elicit grammatical intuitions from subjects, so too can we elicit naive physical intuitions. As Hayes (1985, 31) points out, basic physical intuitions are surprisingly easy to extract from subjects. A theory of grammar will offer an elegant formal scheme within which to derive such intuitions. And a theory of naive physics will likewise, aim at an elegant formal scheme from which the intuitive conclusions follow for some domain. But the worry is the same in each case. Just because we find an elegant theory in which to formally represent or derive the intuitive consequences, why assume that human competence is explained by our internally representing that theory to ourselves?

Consider for example, Hayes's work (1985b) on a naive physics for understanding liquids. The methodology involves attempting a "taxonomy of the possible states liquid can be in" and integrating this with rules about movement, change, and liquid geometry. The propositions and rules discovered and encoded in the theory include a specification of fifteen states of liquid and seventy-four numbered axioms written in predicate calculus. The upshot is a formal system capable of supporting some qualitative reasoning about the behavior of liquids.

The working assumption of the causal interpretation of naive physics is that human competence in this area is due to our having internalized a theory having such axioms. An alternative is to suppose that the axioms offer a formal description of the space of the intuitions we in fact generate in some other way, perhaps by means of a more direct simulation of the properties of liquids in which the formal syntactic elements do not admit of any neat mapping onto entities and relations defined in natural language. This dislocation of the syntax of computational activity from the semantics of natural-language sentences used to describe its products is just what we find in, for example, connectionist systems. And such systems are highly adapted to modeling problem domains involving the simultaneous satisfaction of many soft constraints (e.g., vision). Liquid behavior may well be suited to just such a dynamic model. If so, the intuitions of naive subjects will be a poor clue to the computational structure of the system whose output consists of those very intuitions.

It may be thought that the distinction between description and cause is just a new way of phrasing Marr's well-known distinction between task analysis and algorithm (level 1 versus level 2). This is a mistake. Marr's picture is one in which the details of task analysis provide a structural blueprint for the algorithmic account. On the model suggested, there need be useful no relation between the structure of a task analysis and the underlying algorithm (image here a description of our naive physical competence and a connectionist model of it).

In sum, there is at least as much room to doubt the causal credentials of naive physics as there is to doubt the causal credentials of grammars. In each case we may well be modeling relations between the *products* of brain process and not the brain process itself. Of course, there will be important relations between the two. If we are to guess *how* we do something, we had better know *what* the something is in some detail. But it should at least be controversial to just assume the kind of direct, semantically transparent relations posited by Hayes for naive physics, by Fodor for thought in general, and by Katz for grammar. In general, if a product is describable in a particular, systematic way that ought not to be taken as conclusive evidence for a similarly articulated computational cause in the brain.[5]

8 *Refusing the Syntactic Challenge*

Just to round off this chapter, let me say a word about what Fodor calls intentional realism, i.e., the belief (sic) that beliefs and desires are real and are causes of actions. I suspect that Fodor is driven to defend the position that computational articulation in the brain mirrors the structure of ascriptions of propositional attitudes by a fear that beliefs and desires can only be *causes* if they turn up in formal guise as part of the physical story behind intelligent behavior. But this need not be the case. If belief and desire talk is a holistic net thrown over an entire body of intelligent behavior, we need not expect regular syntactic analogues to particular beliefs and desires to turn up in the head. All we need is that there should be *some* physical, causal story, and that talk of beliefs and desires should make sense of behavior. Such making sense does involve a notion of cause, since beliefs do cause actions. But unless we believe that there is only one model of causation, the physical, this needn't cause any discomfort (see also the argument in the appendix).

Fodor's approach is dangerous. By accepting the bogus challenge to produce syntactic brain analogues to linguistic ascriptions of belief contents, he opens the Pandora's box of eliminative materialism. For if such analogues are not found, he must conclude that there are no beliefs and desires. The mere possibility of such a conclusion is surely an effective *reductio ad absurdum* of any theory that gives it house space.

Chapter 9
Structured Thought, Part 2

1 *Good News and Bad News*

First, the good news. PDP, as illustrated in chapters 5 to 7, affords an approach to computational modeling that should be attractive to anyone engaged in what I have called causal cognitive science. That is, it should be attractive to those who seek to model the in-the-head computational causes of intelligent behavior. Its principal merits include the power of its learning algorithms, its fine-grained shading of meaning, free generalization, and the flexibility that goes with distributed representations of microfeatures.

Now the bad news. PDP, as illustrated in chapters 5 to 7, affords an approach to computational modeling that should be unattractive to anyone engaged in what I have called causal cognitive science. That is, it should be unattractive to those who seek to model the in-the-head computational causes of intelligent behavior. Its principal demerits include the power of its learning algorithms, its fine-grained shading of meaning, free generalization, and the flexibility that goes with distributed representations of microfeatures.

All this is not as contradictory as it sounds. The very properties of PDP models that are advantageous in *some* problem domains are disadvantageous in others, just as being well adapted to survive underwater may be a major disadvantage when beached on dry land.

Hints of such a dark side to PDP were dropped in chapter 7. It is now time to brave the demons. I begin by outlining a particular PDP model in section 2. I then report an influential critique of that model (Pinker and Prince 1988) in section 3 and raise some quite-general worries in section 4. I then give considerable attention to the moral of the story. In the closing section (section 6) I reject Pinker and Prince's moral and put something more ecumenical in its place.

2 *The Past-Tense-Acquisition Network*

The particular PDP model that Pinker and Prince use as the focus of their attack is the past-tense-acquisition network described in Rumelhart and

McClelland 1986 (216–271). The point of the exercise for Rumelhart and McClelland was to provide an alternative to the psychologically realistic interpretation of theories of grammar described briefly in the previous chapter. The counter-claim made by Rumelhart and McClelland is that "the mechanisms that process language and make judgments of grammaticality are constructed in such a way that their performance is characterisable by [grammatical] rules, but that the rules themselves are not written in explicit form anywhere in the mechanism" (1986, 217).

Thus construed, the past-tense-acquisition network, would aim to provide an alternative to what I called propositional psychological realism in chapter 8, section 6, i.e. the view that grammatical rules are encoded in a sentential format and read by some internal mechanism. But this, as we saw, is a very radical claim and is by no means made by all the proponents of conventional symbol-processing models of grammatical competence. It turns out, however, that this PDP model *in fact* constitutes a challenge even to the weaker, and more commonly held, position of structural psychological realism. Structural psychological realism is here the claim that the in-the-head information-processing system underlying grammatical competence is structured in a way that makes the rule-invoking description *exactly* true. As Pinker and Prince put it, "Rules *could* be explicitly inscribed and accessed, but they *also* could be implemented in hardware in such a way that every consequence of the rule-system holds. [If so] there is a clear sense in which the rule-theory is validated" (1988, 168).

The past-tense network challenges structural psychological realism by generating the systematic behavior of past-tense formation without respecting the information-processing articulation of a conventional model. At its most basic, such articulation involves positing separate, rule-based mechanisms for generating the past tense of regular verbs and straightforward memorization mechanisms for generating the past tense of irregular verbs. Call these putative mechanisms the nonlexical and the lexical components respectively. On the proposed PDP model, "The child need not decide whether a verb is regular or irregular. There is no question as to whether the inflected form should be stored directly in the lexicon or derived from more general principles.... A uniform procedure is applied for producing the past tense form in every case" (Rumelhart and McClelland 1986, 267).

One reason for positing the existence of a rule-based, nonlexical component lies in the developmental sequence of the acquisition of past tense competence. It is this developmental data that Rumelhart and McClelland are particularly concerned to explain in a novel way. The data show three stages in the development of a child's ability to correctly generate the past tense of verbs (Kuczaj 1977). In the first stage the child can give the correct form for a small number of verbs, including some regular and some irregular ones. In the second stage the child overregularizes; she seems to have learned the regular "-ed" ending for English past tenses and can give this ending for new and even made-up verbs. But she will now mistakenly give an "-ed" ending for irregular verbs, including ones she got right at stage one. The overregularization stage has two substages, one in which the present form gets the -ed ending (e.g., "come" becomes "comed") and one in which the past form gets it (e.g., "ate" becomes "ated" and "came" becomes "camed"). The third and final stage is when the child finally gets it right, adding "-ed" to regulars and novel verbs and generating various irregular or subregular forms for the rest.

Classical models, as Pinker and Prince note, account for this data in an intuitively obvious way. They posit an initial stage in which the child has effectively memorized a small set of forms in a totally unsystematic and unconnected way. This is stage one. At stage two, according to this story, the child manages to extract a rule covering a large number of cases. But the rule is now mistakenly deployed to generate *all* past tenses. At the final stage this is put right. Now the child uses lexical, memorized, item-indexed resources to handle irregular cases and nonlexical, rule-based resources to handle regular ones.

Classical models, however, typically exhibit a good deal more structure than this bare minimum (see, e.g., the model in Pinker 1984). The processing is decomposed into a set of functional components including a lexicon of structural elements (items like stems, prefixes, suffixes, and past tenses), a structural rule system for such elements, and phonetic elements and rules. A classical model so constructed will posit a variety of mechanisms that represent the data differently (morphological and phonetic representations) with access and feed relations between the mechanisms. In a sense, the classical models here are transparent with respect to the articulation of linguistic theory. Distinct linguistic theories dealing with, e.g., morphology and phonology are paired with distinct in-the-head, information-processing mechanisms.

The PDP model challenges this assumption that in-the-head mechanisms mirror structured, componential, rule-based linguistic theories. It is not necessary to dwell in detail on the Rumelhart and McClelland model to see why this is so. The model takes as input a representation of the verb constructed entirely out of phonetic microfeatures. It uses a standard PDP pattern associator to learn to map phonetic microfeature representations of the root form of verbs to a past-tensed output (again expressed as a set of phonetic microfeatures). It learns these pairings by the usual iterated process of weight adjustments described in previous chapters. The basic structure of the model is thus: phonetic representations of root forms are input into a PDP pattern associator, and phonetic representations of past forms result as output.[1]

The information processing structure of the classical model is thus dissolved. One kind of mechanism is doing all the work both for the regular and irregular forms (recall the quote from Rumelhart and McClelland 1986, 267). And none of the system's computational operations are explicitly defined to deal with such entities as verb stems, prefixes and suffixes (note that this is not just a lack of *labels*; the system nowhere accords any special status to the morphological chunks of words that such labels pick out). As noted by Pinker and Prince, the radical implications of such a model include

- The use of a direct phonetic modification of the root without any abstract morphological representation,
- The elimination of any process dealing specially with lexical items as a locus of idiosyncrasy,
- The use of a qualitatively identical system for regular and irregular occurrences (adapted from Pinker and Prince 1988, 95).

There is thus a quite-extensive dissolution of the structure of a classical model. Not only do we fail to find any explicit tokening of rules such as "add '-ed' to form regular past tenses," but more important, we don't even find any broad articulation of the system into distinct components, one dealing with rule-based behavior and another dealing with exceptional items.

To its undeniable credit the Rumelhart and McClelland model is able to generate much of the required behavior (e.g., the three stages of development) without any such structuring. In so doing it relies on the usual distinctive properties of PDP models, that is, on automatic shading of meaning, blending, and generalization (see chapters 5 to 7). Thus, for example, it finally deals with new cases as if they were regular verbs because this is the correct generalization of the overall thrust of its training input data. The "-ed" ending, we might say, has by then worn down a very deep groove indeed. Nonetheless, the special context provided by inputting a known irregular root can override this groove and cause the correct irregular inflection *but only after sufficient training*. The model thus goes through a stage of overregularizing and learns in time to get it right. Most impressively, the model also produces the second kind of overregularization error observed at stage two: it also overregularizes by adding "-ed" to the *past* tense of irregular verbs, producing errors like "camed," "ated." The explanation of this must lie in the system's *blending* two known patterns from "eat" to "eated" (the regular "-ed" ending) and from "eat" to "ate," and these yield "ated" (see Pinker and Prince 1988).

The PDP model thus recapitulates the three stages of development as follows:

Stage 1. There is simple encoding of a variety of present-past pairings.

Stage 2. The automatic generalization mechanism extracts a regularity implicit in the data and then knows the standard "-ed" ending. For a while this pattern swamps the rest, and causes overregularization. Further training begins to remind the system of the exceptions. But now we find a blend of the "-ed" pattern and the exception patterns, yielding "ated"-type errors.

Stage 3. Further gradual tuning puts it all right. The exceptions and the regular patterns peacefully coexist in a single network.

All of this is just rosy, but darkness looms just around the corner.

3 *The Pinker and Prince Critique*

Pinker and Prince (1988) raise a number of objections to a PDP model of children's acquisition of the past tense. Some of these criticisms are specific to the particular PDP model just discussed, while the others are at least suggestive of difficulties with any nontrivial PDP model of such a skill.[2] I shall only be concerned with difficulties of this last kind. Such cases can be roughly grouped into four types. These concern (1) the model's overreliance on the environment as a source of structure (2) the power of the PDP learning algorithms (this relates to the counterfactual space occupied by such models, a space that is argued to be psychologically unrealistic), (3) the use of the distinctive PDP operation of blending, and (4) the use of microfeature representations.

Overreliance on the environment

The Rumelhart and McClelland model, we saw, made the transition from stage 1 (rote knowledge) to stage 2 (extraction of regularity). But how was this achieved? It was achieved, it seems, by first exposing the network to a population mainly of irregular verbs (10 verbs, 2 regular) and *then* presenting it with a massive influx of regular verbs (410 verbs, 344 regular). This sudden and dramatic influx of regular verbs in the training population is the sole cause of the model's transition from stage one to stage two. Thus, "The model's shift from correct to overregularized forms does not emerge from any endogenous process: it is driven directly by shifts in the input" (Pinker and Prince 1988, 138). By contrast, some developmental psychologists (e.g., Karmiloff-Smith [1987]) believe that the shift is caused by an internally driven attempt to organize and understand the data. Certainly, there is no empirical evidence that a sudden shift in the nature of the input population must precede the transition to stage 2 (see Pinker and Prince, 1988, 142).

The general point here is that PDP models utilize a very powerful learning mechanism that, when given well-chosen inputs, can learn to produce almost any behavior you care to name. But a deep reliance on highly structured inputs may reduce the psychological attractiveness of such models. Moreover, the space of counterfactuals associated with an input-driven model may be psychologically implausible. Given a different set of inputs, these models might go straight to stage 2, or even regress from stage 2 to stage 1. It is at least not obvious that human infants enjoy the same degree of freedom.

The power of the learning algorithms

This is a continuation of the worry just raised. The power of PDP systems to extract statistical regularities in the input data, it is argued, is simply too great to be psychologically realistic. Competent speakers of English can't easily learn the kinds of regularity that a PDP model would find unproblematic. Such a model could learn what Pinker and Prince describe as "the quintessential unlinguistic map relating a string to its mirror-image reversal" (1988, 100). Human beings, it seems, have extreme difficulty learning such regularities. But a good explanation of language acquisition, Pinker and Prince rightly insist, must explain what we *cannot* learn as well as what we can. One way to explain such selective learning capacities is to posit a higher degree of *internal* organization geared to certain kinds of learning. Such organization is found in classical models. The price of dissolving such organization and replacing it with structured input may be a steep reduction in broader psychological plausibility.

Blending

We saw in section 2 above how the model generates errors by blending two such patterns as from "eat" to "ate" and from "eat" to "eated" to produce the pattern from "eat" to "ated." By contrast a conventional rule-based account would posit a mechanism specifically geared to operate on the stems of regular verbs, inflecting them as required. If this nonlexical component were mistakenly given "ate" as a stem, it would simply inflect it sausage-machine fashion into "ated." The choice, then, is between an explanation by blending within a single mechanism and an explanation of misfeeding within a system that has a distinct nonlexical mechanism. Pinker and Prince (1988, 157) point to evidence which favors the latter, classical option.

If blending is the psychological process responsible, it is reasonable to expect a whole class of such errors. For example, we might expect blends of common middle-vowel changes and the "-ed" ending (from "shape" to "shipped" and from "sip" to "sepped"). Children exhibit no such errors. If, on the other hand, the guilty process is misfeed to a nonlexical mechanism, we should expect to find *other* errors of inflection based on a mistaken stem (from "went" to "wenting"). Children do exhibit such errors.

Microfeature representations
The Rumelhart and McClelland model relies on the distinctive PDP device of distributed microfeature representation. The use of such a form of representation buys a certain kind of automatic generalization. But it may not be the right kind. The model, we saw, achieves its ends without applying computational operations to any syntactic entities with a projectible semantics given by such labels as "stem" or "suffix." Instead, its notion of stems is just the center of a state space of instances of strings presented for inflection into the past tense. The lack of a representation of stems as such deprives the system of any means of encoding the *general* idea of a regular past form (i.e., "stem + ed"). Regular forms can be produced just in case the stem in a newly presented case is sufficiently similar to those encountered in training runs. The upshot of this is a much more constrained generalization than that achieved within a classical model, which incorporates a nonlexical component. For the latter would do its work *whatever* we gave it as input. Whether this is good or bad (as far as the psychological realism of the model is concerned) is, I think, an open question. For the moment, I simply note the distinction. (Pinker and Prince clearly hold it to be bad; see Pinker and Prince 1988, 124.)

A more general worry, stemming from the same root, is that generalization based on *pure* microfeature representation is *blind*. Pinker and Prince note that when humans generalize, they typically do so by relying on a theory of which microfeatures are *important* in a given context. This knowledge of salient features can far outweigh any more quantitative notion of similarity based simply on the number of common microfeatures. They write, "To take one example, knowledge of how a set of perceptual features was caused ... can override any generalizations inspired by the object's features themselves: for example, an animal that looks exactly like a skunk will nonetheless be treated as a raccon if one is told that the stripe was painted onto an animal that had raccon parents and raccoon babies" (Pinker and Prince 1988, 177). Human generalization, it seems, is not the same as the automatic generalization according to similarity of microfeatures found in PDP. Rather, it is driven by high-level knowledge of the domain concerned.

To bring this out, it may be worth developing a final example of my own. Consider the process of understanding metaphor, and assume that a successful metaphor illuminates a target domain by means of certain features of the home domain of the metaphor. Suppose further that both the metaphor and the target are each represented as sets of microfeatures thus: $\langle \mathrm{MMF}_1, \ldots, \mathrm{MMF}_n \rangle$ and $\langle \mathrm{TMF}_1, \ldots, \mathrm{TMF}_n \rangle$ (MMF = metaphor microfeature, TMF = target microfeature). It might seem that the necessary capacity to conceive of the target in the terms suggested by the metaphor is just another example of shading meaning according to context, a capacity that as we've seen, PDP systems are admirably suited to exhibit. Thus, just as we earlier saw how to conceive of a bedroom along the lines suggested by inclusion of a sofa, so we might now expect to see how to conceive of a raven along the lines suggested by the contextual inclusion of a writing desk.

But in fact there is a very importance difference. For in shading the meaning of bedroom, the *relevant* microfeatures (i.e., sofa) were already specified. Both the joy and mystery of metaphor lies in the lack of any such specification. It is the job of one who hears the metaphor to *find* the salient features and *then* to shade the target domain accordingly. In other words, we need somehow to fix on a salient subset of $\langle MMF_1, \ldots, MMF_n \rangle$. And such fixation must surely proceed in the light of high-level knowledge concerning the problem at hand and the target domain involved. In short, not all microfeatures are equal, and a good many of our cognitive skills depend on deciding *according to high-level knowledge* which ones to attend to in a given instance.

4 *Pathology*

And the bad news just keeps on coming. Not only do we have the charges of the Pinker and Prince critique to worry about. There is also a body of somewhat recalcitrant pathological data.

Consider the disorder known as developmental dysphasia. Developmental dysphasics are slow at learning to talk, yet appear to suffer from no sensory, environmental, or general intellectual defect. Given the task of repeating a phonological sequence, developmental dysphasics will typically return a syntactically simplified version of the sentence. For example, given "He can't go home," they produce "He no go" or "He no can go." The simplifications often include the loss of grammatical morphemes—suffixes marking tense or number—and generally do not affect word stems. Thus "bees" may become "bee," but "nose" does not become "no." (The above is based on Harris and Coltheart 1986, 111.) The existence of a deficit that can impair the production of the grammatical morphemes while leaving the word stem intact seems prima facie to be evidence for a distinct nonlexical mechanism. We would expect such a deficit whenever the nonlexical mechanism is disengaged or its output ignored for whatever reason.

Or again, consider what is known as surface dyslexia.[3] Some surface dyslexics lose the capacity correctly to read aloud irregular words, while retaining the capacity to pronounce regular words intact. When faced with an irregular words, such patients will generate a regular pronounciation for it. Thus, the irregular word "pint" is pronounced as if it rhymed with regular words like "mint." This is taken to support a dual-route account of reading aloud, i.e., an account in which a nonlexical component deals with regular words. "If the reading system does include these two separate processing components, it might be possible that neurological damage could impair one component whilst leaving the other intact, to produce [this] specific pattern of acquired dyslexia" (Harris and Coltheart 1986, 244). Such data certainly *seems* to support a picture that includes at least some distinct rule-based processing, a picture that on the face of it is ruled out by single-network PDP models.

However, caution is needed. Martin Davies has pointed out that such a conclusion may be based on an unimaginative idea of the ways in which a single network could suffer damage (Davies, forthcoming, 19). Davies does not develop a specific suggestion in print,[4] but we can at least imagine the following kind of case. Imagine a single network in which presented words must yield a certain level of activation of some output units. And imagine that by plugging into an often-repeated pattern, the regular words have, as it were, worn a very deep groove into the system. With sufficient training, the system can *also* learn to give correct outputs (pronounciation instructions) for irregular words. But the depth of groove here is always less than that for the regular words, perhaps just above the outputting threshold. Now imagine a kind of damage that decrements *all* the connectivity strengths by 10 percent. This could move *all* the irregular words below the threshold, while leaving the originally very strong regular pattern functional. This kind of scenario offers at least the beginnings of a single network account of surface dyslexia. For some actual examples of the way PDP models could be used to account for pathological data, see McClelland and Rumelhart 1986, which deals with various amnesic syndromes.

Pathological data, I conclude, at best *suggests* a certain kind of classical structuring of the human information-processing system into lexical and nonlexical components. But we must conclude with Davies that such data is not compelling in advance of a thorough analysis of the kinds of breakdown that complex PDP systems can exhibit. It seems, then, that we are left with the problems raised by the Pinker and Prince critique. In the next section I shall argue that although these problems are real and significant, the conclusions to which they lead Pinker and Prince are by no means commensurate with their content.

5 *And the Moral of the Story Is . . .*

A direct response to the Pinker and Prince criticisms could no doubt be constructed. It could be argued, for example, that what lets a particular PDP network in question down is just its *choice* of microfeatures and that many of the other "general" criticisms flow from that. Thus, it might be that the microfeatures that nature actually focuses on in language acquisition constrain our learning in the ways required, (e.g., by making pairings of words with their mirror-image reversals effectively unlearnable).[5] The lack of certain kinds of blending errors might be explained in the same way. As for fixing on salient features for generalization, perhaps a self-programming network could be constructed that, in a sensitive context-driven way, amends connectivity weights to suit current needs. Similarly, when one natural feature is of great biological importance, we might expect to find a high weight on the activations to which that unit gives rise. Pinker and Prince complain that a network that automatically generalizes by microfeatures couldn't help but treat in the same way two snakes similar in appearance but one poisonous and the other not. But this is simply not true; the weight on the "poisonous" microfeature could be so high as to have dramatic effects whenever that unit is active.

I shall not, however, pursue such lines of response. In the long term, a more-indirect response will be both more effective and more interesting. The indirect response is to grant the general form of Pinker and Prince's worries (shared by many cognitive scientists), but to contest the moral of the story itself.

The worries of section 3 together suggest the need for

(1) More information-processing *structure* in a PDP model of language acquisition, e.g., a morphological as well as a phonetic component,

(2) Some kind of control structure able, e.g., to specify salient microfeatures for inductive generalization,

(3) Some capacity for labeling and variable binding to allow, e.g., the representation of the general idea of a verb stem.

There is, I submit, nothing especially radical here. Many PDP theorists recognize the need for modular or hierarchical organization to satisfy need (1) and control and variable binding to satisfy needs (2) and (3). These demands are all perfectly explicit in Norman 1986, 539–543. To the degree that such facilities turn out to be required for high-level cognitive modeling, there will have to be some broadly classical componential structuring of the information-processing systems concerned.

But it by no means follows that any such model will be a mere implementation of a classical theory. Yet this is exactly what Pinker and Prince lead us to expect. They write, "If the subcomponents of a traditional account were kept distinct in a PDP model, mapping onto distinct subnetworks or pools of units with their own inputs and outputs, or onto distinct layers of a multilayer network, one would naturally say that the network simply implemented the traditional account" (Pinker and Prince 1988, 179). Or again, "Subsymbolism ... will not be indicated if the principal structures of ... hypothetical improved models turn out to be dictated by higher-level theory rather than by micronecessities. To the extent that connectionist models are not mere isotropic node tangles, they will themselves have properties that call out for explanation. We expect that in most cases these explanations will constitute the macro-theory of the rules that the system would be said to implement" (Pinker and Prince 1988, 171). There are two claims here that need to be distinguished.

(1) Any PDP model exhibiting some classical componential structuring is just an implementation of a classical theory.

(2) The *explanation* of this broad structuring will typically involve the use of classical rule-based models.

Claim (1) is clearly false. *Even if* a large connectionist system needs to deploy a complete, virtual, symbol-processing mechanism (recall chapter 7), it by no means follows that the overall system produced merely implements a classical theory of information processing in that domain. This is probably best demonstrated by some examples.

Recall the example (chapter 8, section 4) of a subconceptually implemented rule interpreter. This is a virtual symbol processor—a symbol processor and rule-user realized in a PDP substructure. Now take a task such as the creation of a mathematical proof. In such a case, we saw, the system could use characteristic PDP operations to generate candidate rules that would be passed to the rule interpreter for inspection and deployment. Such a system has the best of both worlds. The PDP operations provide an intuitive (best-match), context-sensitive choice of rules. The classical operations ensure the validity of the rule (blends are not allowed) and its strict deployment.

Some such story could be told for any truly rule-governed domain. Take chess, for example. In such a domain a thoroughly soft and intuitive system would be prone to just the kinds of errors suggested by Pinker and Prince. The fact that someone learns to play chess using pieces of a certain shape ought not to cause her to treat the bishops in a new set as pawns because of their microfeature similarity to the training pawns. Chess constitutes a domain in which absolute hard, functional individuation is called for; it also demands categorical and rigid rule-following. It would be a disaster to allow the microfeature similarity of a pawn to a bishop to prompt a blending of the rules for moving bishops and pawns. A blend of two good rules is almost certain to be a bad one. Yet a *combined* PDP and virtual symbol-processing system would again exhibit all the advantages outlined. It would think up possible moves fluidly and intuitively, but it could subject these ideas to very high-level scrutiny, identify pieces by hard, functional individuation and be absolutely precise in its adherence to the explicit rules of the game.

As a second example, consider the problem of understanding metaphor raised earlier. And now imagine a combined PDP and virtual symbol-processing (VSP) system that operates in the following way. The VSP system inspects the microfeature representation of the metaphor and the target. On the basis of high-level knowledge of the target domain it chooses a salient set of metaphor microfeatures. It then activates that set and allows the characteristic PDP shading process to amend the representation of the target domain as required.

Finally, consider the three-stage-developmental case itself, and imagine that there is, as classical models suggest, a genuine distinction between lexical and nonlexical processing strategies. But suppose, in addition, that the nonlexical process is learned by the child and that the learning process itself is to be given a PDP model. This yields the following picture:

> *Stage 1.* Correct use, unsystematic. This stage is explained by a pure PDP mechanism of storage and recall.
>
> *Transition.* A PDP model involving endogenous (and perhaps innate) structuring, which forces the child to generate a nonlexical processing strategy to explain to itself the regularities in its own language production
>
> *Stage 2.* Overregularization due to sudden reliance on a newly formed nonlexical strategy
>
> *Transition.* A PDP model of tuning by correction
>
> *Stage 3.* Normal use. The coexistence of a pure PDP mechanism of lexical access and a nonlexical mechanism implemented with PDP

If some such model were accurate (and something like this model is in fact contemplated in Karmiloff-Smith 1987), we would *not* have a classical picture of development, although we might have a classical picture of adult use.[6]

To sum up, the mere fact that a system exhibits a degree of classical structuring into various components (one of which might be a rule interpreter) does not force the conclusion that it is a mere implementation of a classical theory. This is so because (a) the classical components may call and access powerful PDP operations of matching, search, blending and generalization and (b) the developmental process by which the system achieves such structure may itself require a PDP explanation. Claim (1) thus fails. It may be, however, that to understand *why* the final system must have the structure it does, we will need to think in classical, symbol-manipulating terms. This second claim (claim 2, p. 171) is considered in the next section.

6 *The Theoretical Analysis of Mixed Models*

What are the theoretical implications of mixed PDP and VSP models? One thought, which we have already rejected, is that any such model must be a mere implementation of classical information processing in the domain. An equally radical and equally misguided thought is that classical models in such cases are at best approximations to the true story told at the level of a PDP analysis of units and passing values. The mistake in both cases is to think in terms of one model for each task. For as we saw in chapter 7 section 6, we individuate tasks according to our particular interests. But our performance of any top-level task may need to be computationally explained in terms of a number of interacting virtual machines. And some of these virtual machines may need to be understood in classical terms, e.g., in terms of their performing classical operations on items such as verb stems, suffixes, numbers, phonemes, morphemes, English words, and so on. And others may require understanding in connectionist terms, e.g., in terms of operations on microfeatures that are not semantically transparent. Thus, recall Fodor's argument in chapter 8. It is debatable, but perhaps these arguments succeed in demonstrating the need for *some* computational operations defined over sentential items (and hence semantically transparent items). This seems plausible in the case of what he calls "causal trains of thought," conscious sequences of sententially formulated mental states (see Fodor 1987, 147). But it would hardly follow from this that all or even most of our cognitive activity is best explained in such terms.

When a single task is carried out by a complex of virtual machines all implemented in a PDP architecture but some simulating classical operations of hard matching and serial processing, a good psychological model of the information processing involved will have to be *multiplex*. Recall Smolensky's idea of a mathematical proof generator that has an intuitive component and a classical component. In that case, the process of finding candidate rules to apply involves best-match operations requiring a PDP explanation. But the final selection and deployment of the candidate rules require explanation in terms of classical operations. Moreover, as claim 2 implies, this *overall* setup itself needs to be understood as a response to the genuine, hard, rule-based nature of mathematical proof.

Or again, recall the mixed model of metaphor understanding. A theoretical analysis of the system requires a classical account of the *choice* of salient microfeatures (including errors in choice), a PDP account of shading the target domain, and a PDP account of the distributed microfeature representation of target and metaphor. Moreover, mixed systems will be susceptible to various kinds of breakdown. Some breakdown patterns may be explicable *only* by adverting to the underlying PDP substrate used to implement a symbol processor; others may make sense at the level of virtual symbol-processing operations; and still others may affect the pure PDP component itself. We might speculate that breakdowns of this last kind (the loss of some pure PDP processing power that leaves virtual symbol-processing capacities intact) are behind some of the fascinating cases reported in Sacks (1986).[7]

Mixed models thus require multiplex forms of psychological or computational explanation. Not just different cognitive *tasks*, but different aspects of the *same* task now seem to need different kinds of algorithmic explanation. Since humans must frequently negotiate some truly rule-governed problem domains (e.g., chess, language, mathematics), some form of mixed model may well be the most effective explanation. The apparent success of thoroughly soft PDP systems in negotiating some such domains (e.g., the model of past-tense acquisition) may be due to a concealed bolt-on symbol-processing unit, us. In the model of past-tense acquisition the system received stems and then inflected versions because we chose to divide the verbs up like that. Pinker and Prince describe this choice as relying on "intuitive protolinguistics." So in that sense, even the Rumelhart and McClelland system has a bolt-on symbolic component. At any rate, if mixed models are required (for whatever reason), the consequences must include the general failure of the uniformity principle (see chapter 7). More specifically, they must include:

- The rejection of the claim that any model exhibiting classical componential structure is a mere implementation of a classical theory,
- The rejection of the claim that any classical account is at best an approximation to a correct PDP-based account.

Instead, correct explanations must be geared to the virtual machine responsible for particular aspects of performing the task. All of this is nicely ecumenical, I'm sure.

It would be boring, however, to close without making at least one inflammatory claim. The *power* behind our gross symbol-processing capacities—a factor that makes us thinkers and, e.g., BACON not—may well be the subsymbolic, pattern-matching power of something like a PDP mechanism operating within us. There is a strong intuition that manipulating gross symbolic structures models the *form* of some of our thought but somehow leaves out the content. The intuition is often put by saying that such programs have no understanding of what the symbol manipulations mean. Perhaps, then, understanding involves spontaneously seeing patterns, spotting similarities, shading meanings, and so on. (This position is most strongly advanced in Hofstadter 1985.) Of the two modes of thought treated in this book, it would seem the PDP mode is in some sense primary. This certainly fits our normal usage. Many of us allow that lower animals have thoughts of some kind. They are plausibly seen as advanced, complex PDP machines that have not yet developed our capacities for symbolic representation. Yet we deny thoughts to BACON and SHRDLU, programs that certainly manipulate gross symbolic representations but lack any rich pattern matching substructure.

If this picture is correct, we should maintain a dual thesis concerning explanation and instantiation. We should hold that good psychological *explanations* will often involve mixed models and hence will require analysis in both PDP *and* classical symbol-manipulating terms. But we may *also* hold that instantiating any contentful psychological state requires not *just* the manipulation of gross symbolic structures but also access to the output of a powerful subsymbolic processor. A virtual symbol processor provides guidance and rigor; the PDP substrate provides the fluidity and inspiration without which symbol processing is but an empty shell. In words that Kant never used: subsymbolic processing without symbolic guidance is blind; symbolic processing without subsymbolic support is empty.

Chapter 10
Reassembling the Jigsaw

1 The Pieces

All the pieces of the jigsaw are now before us, and their subgroupings are largely complete. Semantically transparent AI models have been described and compared with highly distributed connectionist systems. Various worries about the power and methodology of both kinds of work have been presented. The possibility of mixed models of cognitive processing has been raised, and the nature of folk-psychological talk and its role in a science of cognitive processing has been discussed. Along the way I have criticized the arguments in favor of Fodor's radical cognitivism, and I was forced to distinguish two projects within cognitive science: one descriptive and involving the essential use of classical representations; the other concerned with modeling the computational *causes* of intelligent behavior and typically *not* dependent on such representations. At the end of the previous chapter I also drew a distinction *internal* to causal cognitive science: the distinction between the project of psychological *explanation* (laying out the computational causes of intelligent behavior) and that of *instantiation* (making a machine that actually has thoughts). These two projects, I suggested, may come apart. This final chapter (which also functions as a kind of selective summary and conclusion) expands on this last piece of the jigsaw and tries to display as clearly as possible the overall structure of what I have assembled. In effect, it displays a picture of the relations of various parts of an intellectual map of the mind.

One word of warning. Since I should be as precise as possible about what each part of this intellectual map is doing, for the duration of the chapter I shall largely do away with shorthand talk of representations, beliefs, and so on to describe contemporary computer models (recall chapter 6, section 2, and chapter 5, footnote 4). At times this will result in language that is somewhat cumbersome and drawn out.

2 *Building a Thinker*

What does it take to build a thinker? Some philosophers are sceptical that a sufficient condition of being a thinker is satisfying a certain kind of formal description (see chapter 2). Such worries have typically focused on the *kinds* of formal descriptions appropriate to semantically transparent AI. In one sense we have seen virtue in such worries.[1] It has indeed begun to seem that satisfying *certain* formal descriptions is vastly inadequate to ensure that the creature satisfying the description has a cognitive apparatus organized in a way capable of supporting the rich, flexible actual and counterfactual behavior that warrants an ascription of mental states to it. (Apologies for the lengthy formulation—you were warned!) Some reasons for thinking this were developed in chapter 6, where I discussed the holism and flexibility achieved by systems that use distributed representations and superpositional storage.

In short, many worries can usefully be targeted in what I am calling the project of *instantiation.* They can be recast as worries to the effect that satisfying the *kind* of formal description that specifies a conventional, semantically transparent program will never isolate a class of physical mechanisms capable of supporting the rich, flexible *actual and counterfactual* behavior that warrants ascribing mental states to the system instantiating such mechanisms. The first stage in an account of instantiation thus involves the description of the general structure of a mechanism capable of supporting such rich and flexible behavior *at the greatest possible level of abstraction from particular physical devices.* Searle seems to believe that we reach this level of abstraction *before* we leave the realms of biological description (see chapter 2). I see no reason to believe this, although it could conceivably turn out to be true. Instead, my belief is that some nonbiological, microfunctional description, such as that offered by a value-passing PDP approach, will turn out to specify at least one class of physical mechanisms capable of supporting just the kind of rich and flexible behavior that warrants ascribing mental states.

This is not to say, however, that its *merely* satisfying some appropriate formal description will warrant calling something a thinker. Instead, we need to imagine a set of conditions *jointly sufficient* for instantiating mental states, one of which will involve satisfying some microfunctional description like those offered by PDP. I spoke of systems that could be properly credited with mental states if they instantiated such descriptions. And I spoke also of a mechanism that, suitably embodied, connected and located in a system would allow us to properly describe it in mentalistic terms. These provisos indicate the second and final stage of an account of instantiation.

Instantiating a mental state may not be a matter of possessing a certain *internal* structure alone. In previous chapters we discovered two reasons to believe that the configuration of the external world might figure among the conditions of occupying a mental state. The first reason was that the ascription of mental states may involve the world (chapter 3). The content of a belief may vary according to the configuration of the world (recall the twin earth cases [chapter 3, section 4]). And some beliefs (e.g., those involving demonstratives) may be simply unavailable in the absence of their objects. What this suggests is that instantiating *certain* mental states may involve being suitably located and connected to the world. (It does not follow, as far as I can see, that a brain in a vat can have *no* thoughts at all.)

We also noted in chapters 4 and 7 a second way in which external facts may affect the capacity of a system to instantiate mental states. This is the much more practical dimension of exploitation. A system (i.e., a brain or a PDP machine) may need to use external structures and bodily operations on such structures to augment and even qualitatively alter its own processing powers. Thus, suppose that we accept that stage one of an instantiation account (an account of brain structures) involves a microfunctional specification of something like a PDP system. We might *also* hold that instantiating some mental states (for example, all those involving conscious, symbolic and logical reasoning) requires that such systems emulate a different architecture. And we might believe that such emulation is made possible only by the capacity of an embodied system located in a suitable environment to exploit real-world structures to reduce complex, serial processing tasks to an iterated series of PDP operations. PDP systems are essentially learning devices, and learning devices (e.g., babies) come to occupy mental states by interacting with a rich and varied environment. For these very practical reasons the project of full instantiation may be as dependent on embodiment and environmental structure as on internal structure.

Most important of all, I suspect, is the holistic nature of thought ascription. Thoughts, we may say, just *are* what gets ascribed using sentences expressing propositional attitudes of belief, desire, and the like. Such ascriptions are made on the basis of whole nexuses of actual behavior. If this is the case, to have a certain thought is to engage in a whole range of behaviors, a range that, for daily purposes, is usefully codified and explained by a holistically intertwined set of ascribed beliefs and desires. Since there will be no neat one-to-one mapping of thoughts so ascribed to computational brain states (see chapters 3 and 8 especially), it follows *a fortiori* that there will be no computational brain state that is a sufficient condition of having that thought. The project I have called *descriptive* cognitive science in effect gives a formal model of the internal relations of sentences used to ascribe such thoughts. This is a useful project, but instantiating that kind of formal description certainly *won't* give you a thinker. For the sentences merely describe regularities in the behavior and are not geared to pick out the syntactic entities that are computationally manipulated to *produce* the behavior.

To sum up, the project of instantiation is just the project of creating a system properly described as occupying mental states. And it involves two stages (to be pursued cooperatively, not in series). Stage 1 is the description, at the highest possible level of abstraction, of a class of mechanisms capable of supporting the kind of rich, flexible, actual and counterfactual behavior needed to warrant the use of a mentalistic vocabulary. That level of description will turn out, I believe, to be a microfunctional one. It may well turn out to involve in part a microfunctional specification of PDP systems in terms of value passing, thresholds, and connectivity strengths. I also urged that no highly semantically transparent model can fulfil the requirements of stage 1 of an instantiation, despite the claim made by Newell and Simon that such approaches capture the necessary and sufficient conditions of intelligent action (see chapter 1, sections 2, 3, and 4). Stage 2 of the project of instantiation will involve the embodying and environmental embedding of mechanisms picked out in stage 1. Only once these systems are embodied, up, and running in a suitably rich environment will we be properly warranted in our ascriptions of mental states.

3 *Explaining a Thinker*

The project of instantiation and the project of psychological modeling and explanation are different. This may seem obvious, but I suspect a great deal of confusion within cognitive science is a direct result of not attending to this distinction.

First and most obviously, the project of instantiation requires only that we delimit a class of mechanisms capable of providing the causal substructure to ground rich and varied behavior of the kind warranting the ascription of mental states. There may be *many* such classes of mechanisms, and an instantiation project may thus succeed without first delimiting the class of mechanisms that human brains belong to. But we may put this notion aside at least as far as our interest in PDP is concerned. PDP is certainly neurally inspired and aims to increase our knowledge of the class of mechanisms to which we ourselves are in some significant way related.

Second and more importantly, even if the microfunctional description that (for the instantiation project) delimits the class of mechanisms to which we belong is entirely specified by a PDP-style account, correct psychological models and explanations of our thought may *also* require accounts couched at many different levels. To bring this out, recall my account of Marr's picture of the levels of understanding of an information processing task (chapter 1, section 5). *Psychological* explanation, according to Rumelhart and McClelland (1986, 122–124) is "committed to an elucidation of the algorithmic level," i.e., Marr's level 2. For the story at this level—the level that specifies the mode of representation and actual processing steps—provides the explanation of such phenomena as speed, efficiency, relative ease in solving various problems, and graceful degradation (performance with noise, inadequate data, or damaged hardware). That is, the story at the algorithmic level provides the explanation of the performance data with which real psychology is typically interested.

Suppose we accept this broad characterization of the level of psychological interest. *It will not follow that in general, a single computational model serves to explain all such data relating to the performance of a given task.* One reason for this has directly to do with the notion of virtual machines. Thus, imagine a PDP system engaged in the full or partial simulation of a more conventional processor (e.g., the environment manipulator for full simulation and the mathematical prover for partial simulation). In such cases we will need to advert to at least two algorithmic descriptions of the system to explain various kinds of data. The relative ease with which the system solves various problems and the nature of the transformations of representations involved will often require an account couched in terms of the top-level virtual machine, e.g., a production system or a list processor. But speed and graceful degradation will need to be explained by adverting to an algorithmic description of the PDP implementation of some of the functions found in the top level virtual machine. The thought is that not only may different *tasks* require different forms of computational explanation, but different kinds of data pertaining to a single task may likewise require several types of computational models.

At this point someone might object as follows. It may be *convenient* to use a classical serial model at times. But because of the underlying PDP implementation, a full and correct psychological explanation always can in principle be given in PDP algorithms alone.

This is a general reductionist argument that in the extreme is sometimes thought to threaten the integrity of the entire project of psychological explanation. But the specter of reduction need not be feared. For explanation is not *just* a matter of showing a structure that is sufficient to induce or constitute a certain higher-level state or process. It is also a matter of depicting the structure *at the right level.* And the right level here is determined by the need to capture *generalizations* about the phenomena picked out by the science in question. The general point has been made often enough (see, e.g., Pylyshyn 1986, chapter 1), and I shall not labor it here. Instead, I shall merely sketch the relevant instances. Consider the cases of full simulation from chapter 7. Here we have (by hypothesis) a PDP substructure supporting the mode of representation of the input and output and the processing steps and pattern-matching characteristics of a regular

Von Neumann processor running a semantically transparent program. Other computational substructures (e.g., a real Von Neumann machine) can certainly support the same features. And it is precisely these features that determine *some* of the psychologically relevant performance data, such as the relative ease of solving various problems (which is keyed to the mode of representing the problem).

Suppose we tried to give a psychological explanation of this aspect of the performance of the system, using only the formal apparatus of a PDP specification. We would merely succeed in obscuring the fundamental psychological *similarity* of the systems built out of the various substructures adverted to above. Of course, where the performance differs (e.g., in style of degradation and speed), we will need a psychological explanation to explain the difference also. So for those other aspects we may indeed need an algorithmic specification at the PDP level.

The basic point is well expressed by Hilary Putnam when he writes, "Explanation is not transitive" (1981, 207). What *explains* the simulation in any given case need not itself be a good explanation of what the simulation explains if we are to retain the requisite degree of generality.

We can give essentially the same treatment to approximations by conventional accounts (see chapter 6, section 3). Insofar as a particular system *nonaccidentally* approximates the behavior of some other (e.g., a PDP system that behaves largely as if it had the syntactic substructures described in a conventional model), it may be said to *partially simulate* the other system. Now imagine a range of systems all with different formal substructures but all of which nonaccidentally partially simulate the behavior predicted by a conventional model. And suppose, moreover, that the range of cases for which the conventional model is accurate is the same in all of these. In that case, I am inclined to say, there is a genuine psychological generalization that is in need of explanation and that would evade our grasp *unless* we avail ourselves of the model provided in the conventional account.

In general, then, my claim is that the notion of a single formal algorithmic level appropriate to psychological explanation is misplaced. For different tasks and different aspects of the same task may positively *require* a variety of algorithmic models detailing the processing undertaken by various virtual machines. If so, then *some* psychological explanation is properly given at the level of conventional, semantically transparent serial programs. But other phenomena (e.g., creative leaps, flashes of insight, jokes coming to us, analogical understanding, perception, fast expert problem solving, and so forth) seem to require psychological models at the level of PDP (or microfunctional) accounts. Various neuropathological data may also require explanation at this level. Thus, the project of psychological explanation may involve the construction both of microfunctional, PDP accounts and in some cases the construction of serial, symbol-processing accounts.

4 Some Caveats

Part of my project is to assess the importance and role of PDP models in understanding the human mind. And the conclusion seems to be that such models have a major part to play in each of the two projects (instantiation and explanation) just distinguished. Such a conclusion, however, needs to be qualified in at least two regards.

First, PDP mechanisms may turn out to be just one among many kinds of mechanisms capable of supporting the rich, flexible actual and counterfactual behavior demanded of a genuine cognizer. Thus, even if semantically transparent approaches lack the capacity to ground such behavior (as I suspect), it does not follow that *all* thought requires a PDP substrate.

Second, the particular algorithms currently being explored by PDP theorists are almost certainly still inadequate to the task. The brain seems to employ many kinds of parallel cooperative networks, using different kinds of units and connectivity patterns. And this variety may be essential to its power but is not yet present in PDP work, which uses a simple, idealized neuronlike unit. In a recent article on the computation of motion, two leading theorists comment, "Nerve cells exhibit a variety of information-processing mechanisms; the nerve cell membrane produces and propagates many different types of electrical signals." "One can think of the McCulloch and Pitts model [see chapter 5 above] as equating a neuron with a single transistor whereas our model suggests that neurons are more like computer chips with hundreds of transistors each" (Poggio and Koch 1987, 42 and 48). The idealized neurons of current PDP models, it is fair to say, are only a little finer and exhibit only a little more variety than the original McCulloch and Pitts versions. Current models, then, may suffer many severe limitations until workers in the field are in a position to introduce greater detail and variety.

In a similar vein, the learning algorithms currently in favor (the generalized delta rule and Boltzmann machine learning procedure) are most probably inadequate in various ways. For example, as Rudi Lutz has pointed out, the generalized delta rule requires us in effect to tell the machine when it is to *learn* and when it is simply to behave on the basis of what it already knows. Yet this explicit switching of modes is quite counterintuitive as part of any psychological model of human learning.

In short, the very broad brushstrokes of PDP, the general idea of a parallel, value-passing architecture encoding information in distributed patterns of activity and connectivity, will probably constitute its positive contribution to understanding the mind, *not* the particular algorithms and idealized neurons currently under study. (This is not, of course, any criticism of the current work; creating such models and then trying to understand their limits is the best way of improving upon them.)

What the broad brushstrokes of PDP give us is an indication of one way in which a physical, computational mechanism may satisfy the range of constraints on biological cognition developed in previous chapters. The potential capacity of PDP to satisfy these constraints ensures its continued philosophical and psychological interest. Such constraints were seen to include:

- Robustness (a tolerance of local hardware damage),
- Fast sensory processing,
- Sensible action when given partial or inconsistent data,
- Economy of storage and retrieval,
- A capacity to deal with unanticipated situations (e.g., to generalize along unexpected dimensions),
- General flexibility in the use and recovery of stored data,
- A powerful learning capacity,
- Rule-describable behavior without explicit, fixed rules,
- Continuity with the kind of architectures dictated in evolutionarily basic cases (i.e., the constraints imposed by the gradualistic holism of evolutionary change),
- The capacity to shade meanings according to context, to create schemata on demand, and so on.

These constraints and capacities form an interlinked, often overlapping set. Taken together, they amount to a demand for a computational substructure that supports maximally plastic and adaptable behavior while simultaneously sustaining the extraction and storage of regularities and similarities in its input. The capacity of a PDP architecture to satisfy both needs in a fast, natural, and economical way is a stunning achievement. In doing so, it suggests for the first time just how a physical, computational mechanism might support the sensible, flexible, open-ended behavior that philosophers have rightly demanded of any system that warrants a description in a mentalistic vocabulary.

Epilogue
The Parable of the High-Level Architect

This little story won't make much sense unless it is read in the context provided by chapter 4, section 5, and with an eye to the general distinction between semantically transparent and semantically opaque systems.[1]

One fine day, a high-level architect was idly musing (reciting Wordsworth) in the cloistered confines of Kings College Chapel. Eyes raised to that magnificant ceiling, she recited its well-publicized virtues ("that branching roof, self-poised and scooped into ten thousand cells, where light and shade repose . . ."). But her musings were rudely interrupted.

From a far corner, wherein the fabric of reality was oh so gently parting, a hypnotic voice commanded: "High Level Architect, look you well upon the splendours of this chapel roof. Mark well its regular pattern. Marvel at the star shapes decorated with rose and portcullis. And marvel all the more as I tell you, *there is no magic here.* All you see is complex, physical architecture such as you yourself might re-create. Make this your project: go and build for me a roof as splendid as the one you see before you."

The high-level architect obeyed the call. Alone in her fine glass and steel office, she reflected on the qualities of the roof she was to re-create. Above all, she recalled those star shapes, so geometric, so perfect, the vehicle of the rose and portcullis design itself. "Those shapes," she concluded, "merit detailed attention. Further observation is called for. I shall return to the chapel."

There ensued some days of patient observation and measurement. At the end of this time the architect had at her command a set of rules to locate and structure the shapes in just the way observed. These rules, she felt sure, must have been followed by the original designer. Here is a small extract from the high-level architect's notebook.

```
! To create ceiling shapes instruct the builder (Christopher Paul Ewe?)
! as follows:
if (build-shapes) then
  [(space-shapes (3-foot intervals)),
   (align-shapes (horizontal)),
```

(arrange-shapes (point-to-point)),
(locate-shapes (intersection-of-pillar-diagonals))].

Later she would turn her attention to the pillars, but that could wait. When the time came, she felt, some more rules would do the trick. She had an idea of one already. It went "If (locate-pillar) then (make-pillar (45°, star-shape)). It was a bit rough, but it could no doubt be refined. And of course, there'd be lots more rules to discover. "I do hope," she laughed, "that Christopher Paul Ewe is able to follow all this. He'll need to be a fine logical thinker to do so." One thought, however, kept on returning like a bad subroutine. "Why are things arranged in just that way? Why not have some star shapes spaced further apart? Why not have some in a circle instead of in line? Just think of all the counterfactual possibilities. What an unimaginative soul the original architect must have been after all."

Fortunately for our heroine's career, this heresy was kept largely to herself. The Society for the Examination and Reconstitution of Chapels gave her large research grants and the project of building a duplicate ceiling went on. At last a prototype was ready. It was not perfect, and the light and shadow had a subtly different feel to it. But perhaps that was mere supersition. C. P. Ewe had worked well and followed instructions to the letter. The fruits of their labors were truly impressive.

One day, however, a strange and terrible thing happened. An earthquake (unusual for the locale) devastated the original chapel. Amateur video, miraculously preserved, records the event. The high-level architect, upon viewing the horror, was surprised to notice that the star shapes fell and smashed in perfect coincidence with the sway and fall of neighbouring pillars. "How strange," she thought; "I have obviously been missing a certain underlying unity of structure here." The next day she added a new rule to her already massive notebooks: "If (pillar-falls) then (make-fall (neighboring-star-shape))." "Of course," the architect admitted, "such a rule is not *easy* for the builder to follow. But it's nothing a motion sensor and some dynamite can't handle."

Appendix
Beyond Eliminativism

1 *A Distributed Argument*

The main body of the text contains, in a somewhat distributed array, an argument against the use of connectionist models to support the position known as eliminative materialism. In this appendix, I gather those distributed threads and weave them into an explicit rejection of eliminativism. The appendix expands on various hints given in chapters 3 and 10 and connects these, in a somewhat unexpected way, with the idea of mixed symbolic and connectionist models, introduced in chapter 7. As a bonus, it introduces a new and interesting way of describing connectionist systems with a statistical technique known as cluster analysis.

The appendix begins by laying out various types of descriptions of connectionist systems of the kind we have been considering (section 2). In section 3 it goes on to expand on the idea (chapter 10) of explanations that seek to group systems into equivalence classes defined for various purposes. Each such grouping requires a special vocabulary, and the constructs of any given vocabulary are legitimate just insofar as the grouping is interesting and useful. Section 4 then shows that relative to such a model of explanation, the constructs of both symbolic AI and commonsense psychology may have a legitimate role to play in giving psychological explanations. This role is *not* just that of a useful approximation. Section 5 is a speculative section in which the argument for the theoretical usefulness of such symbolic constructs is extended to individual processing in a very natural way. Here the cognizer, in the process of regulating, debugging, and understanding her own representations, creates symbols to stand for sets of distributed activity patterns. The section points out the difficulties for a pure distributed approach that may be eased by the addition of such symbolic constructs, and it relates my speculations to the continuing debate over the "correct" architecture of cognition.

2 Levels of Description of Connectionist Systems

Connectionist systems, like everything else, can be described at a variety of levels, each with its own characteristic vocabulary. The central concern of this appendix lies with the status of various high-level descriptions of such systems. Low-level descriptions include

(1) The numerical specification of weights and activation-passing rules and

(2) Subsymbolic interpretations of the activity of processing units.

High-level descriptions include

(3) The partitioning trees created by performing a cluster analysis on a network,

(4) Descriptions that use the constructs of classical AI (e.g., "schema," "production," and so on), and

(5) The ordinary conceptual-level descriptions of commonsense belief and desire psychology.

Paul Smolensky (1988) has recently attempted the essential work of detailing the status and interrelations of these levels of description. Smolensky's picture (which only applies to what I shall call pure distributed connectionism) is, I believe, technically accurate and represents a major achievement in sketching the theoretical foundations of PDP. But it invites a certain distortion of the role and status of the high-level descriptions—an invitation that the eliminativists cannot refuse. In the remainder of this section I sketch the received attitude to each of the five levels of description.

Level 1, the numerical level

The most precise characterization of the actual processing of a particular connectionist network is mathematical in nature. Such networks, we saw, consist of interconnected units. The connections are weighted and the units are miniprocessors that receive and pass on activation according to mathematical specifications. Thus, the theorist can give a precise characterization of the state of such a system at a particular time by stating a vector of numerical values. Each element in the vector will correspond to the activation value of a single unit. Likewise, it is possible to specify the evolving behavior of a system by an "activation-evolution equation." This is a differential equation that fixes the dynamics of the network. If, as is generally the case, the network is set up to learn, then it will be necessary to specify the dynamics of its learning behavior. This is done by means of another differential equation, the "connection-evolution equation." Such specifications give a complete mathematical picture of the activation and processing profile of any given network. (For a more detailed account see Smolensky 1988, sections 1 and 2.)

These mathematical specifications play a large and important role in connectionist cognitive science. They are often the only way to understand the distinctive wrinkles in the learning behavior of different kinds of connectionist systems (for example, Boltzmann-machine learning versus various forms of supervised learning). They also figure in explanations of specific behaviors and pathologies. In this sense, as Smolensky observes, "the explanations of behavior provided are like those traditional in the physical sciences, unlike the explanations provided by symbolic models" (Smolensky 1988, 1).

Level 2, the subsymbolic level

For all that, however, Smolensky seems especially fond of a slightly higher level of analysis that he calls the subconceptual (or subsymbolic) level. It is at that level, and *not* at the numerical (or mathematical) one, that we find the "complete formal account of cognition." He writes, "Complete, formal and precise descriptions of the intuitive (i.e. connectionist) processor are generally tractable not at the conceptual level, but only at the subconceptual level" (Smolensky 1988, 6–7). But there is, in fact, no inconsistency here. For Smolensky views the subsymbolic as just the semantic (microsemantic) description of the syntactic (units and activation) profile of level 1. He is thus committed to the semantic interpretability of the numerical variables specifying unit activations. This interpretation takes the form of specifying the subsymbolic (or microfeatural) content to which the unit activation corresponds in the context of a particular activation vector. Hence, "the name 'subsymbolic paradigm' is intended to suggest cognitive descriptions built up of *constituents* of the symbols used in the symbolic paradigm; these fine-grained constituents could be called subsymbols and they are the activities of individual processing units in connectionist networks" (Smolensky 1988, 3).

The semantic shift from symbolic to subsymbolic specification is one of the most important and distinctive features of the connectionist approach to cognitive modeling. It is also one of the most problematic. One immediate question concerns the nature of a subsymbol. The level of description at issue is clearly meant to be a level that ascribes content, a level that rather precisely *interprets* the numerical specification of an activation vector by associating the activation of each unit with a content. In an activation vector that amounts to a distributed representation of coffee, we saw how the activation of a single unit may represent such features as hot liquid, burnt odor, and so on. Such examples make it seem as if a subsymbolic feature (or microfeature) is just a partial description, in ordinary-language terms, of the top-level entity in question (coffee). This is certainly the case in most (or perhaps even all) of the toy examples found in the literature. Nonetheless, there is clearly a theoretical commitment to something more radical. Thus, concerning the coffee example, Smolensky adds, "we should really use subconceptual features, but even these features (e.g., 'hot liquid') are sufficiently low level to make the point" (1988, 16).

The official line on the semantic shift (or dimension shift) in connectionist representation is that dimension-shifted representations must be of features that are more subtle than those in an ordinary task analysis of the problem. The claim is that the elements of a subsymbolic program do not refer to the "same concepts as are used to consciously conceptualise the task domain" (Smolensky 1988, 5). Or again, "the units do not have the same semantics as words of natural language" (Smolensky 1988, 6). We can now see that these claims can be taken in two ways. The stronger way is to take the claims to mean that the content to be associated with the activation of a given unit in context *cannot* be captured by any formulation in natural language, however long and hyphenated. The weaker way is to take the claims to mean that individual unit activations don't have the semantics of the single words that occur in a conscious task analysis of the domain. The latter is clearly the safer claim, at least as long as we avoid being too imaginative concerning the nature of the task analysis. But it seems that Smolensky believes that the former, more radical reading will ultimately prove correct. He thus notes, "Semantically, the subconceptual level seems at present rather close to the conceptual level" (1988, 8). But this, he conjectures, is probably because the choice of input and output representations, a crucial factor in determining what a system will learn, is heavily based on "existing theoretical analyses of the domain." It may well be that truly subsymbolic models (i.e., in the strong sense) will not become available unless input and output representations can be divorced from our existing analyses of the domain. Whether this is possible is a question that would take us too far afield.

It is likely, however, that the real importance of subsymbolic representation lies not just in *what* gets represented but in the special properties of the representational medium that connectionists employ. Much work in ordinary AI (vision, natural-language processing) depends, after all, on the representation and manipulation of features quite invisible to daily, conscious reflection on the task at hand. Where the two paradigms differ most radically is surely in the general *mode* of representation and its associated properties. In particular, subsymbolic (i.e., connectionist) representation naturally embodies a kind of *semantic metric* (I owe this term to Andler 1988), which powers the distinctive features of generalization, graceful degradation, and so on. The semantic metric is best pictured as a spatial arrangement of units in a multidimensional space arranged so that semantically related items are coded for by spatially related feature units. This fact renders each individual unit pretty well expendable, since its near neighbors will do *almost* the same job in generating patterns of activation. And it is this same fact that allows such systems to generalize (by grouping the semantically common parts of various items of knowledge), to extract prototypes, and so on. Classical representation does not involve any such built-in notion of semantic metric.

Distributed (i.e., microfeatural) representations with a built-in semantic metric are also responsible for the context dependence of connectionist representations of concepts. Recall that in what I am calling pure distributed connectionism there are no units that represent classical conceptual-level features, such as coffee. Instead, coffee is represented as a set of active microfeatures. The point about context dependence is that this set will *vary* according to the surrounding context. For example, "coffee in cup" may involve a distributed representation of coffee that includes contacting procelain as a microfeature. But "coffee in jar" would not. Conceptual-level entities (or "symbols," to fall in with a misleading terminology) thus have no stable and recurrent analogue as a set of unit activations. Instead, the unit-activation vector will vary according to the context in which the symbol occurred. This, we saw, is an important feature (though *at times* it may be a positive defect). It is directly responsible for the oft-cited fluidity of connectionist representation and reasoning.

If it is not the dimension shift *in itself* so much as the dimension shift in conjunction with a built-in semantic metric that is the crucial fact in connectionist processing, then a question arises about the status of the subsymbolic level of description. For such descriptions seemed to involve just listing a set of microfeatures corresponding to an activation vector. But such a listing leaves out all the facts of the place of each feature in the general metric embodied by the network. And these facts seem to be of great semantic significance. What a microfeature means is not separable from its place in relation to all the other representations the system embodies. For this reason, I would dispute the claim that subsymbolic description (at least, if it is just a listing of microfeatures) affords an accurate interpretation of the full numerical specifications available in level 1. Perhaps the resources of natural language (however cannily deployed) are in principle incapable of yielding an accurate interpretation of an activation vector. At first sight, such a concession may seem to give the eliminativist an easy victory. Fortunately, this impression is wrong, as we shall see in due course.

Level 3, cluster analysis

This level of analysis does not appear in Smolensky's treatment. Rather, it occurs as part of the methodology developed by Rosenberg and Sejnowski for the analysis of NETtalk. (For an account of NETtalk though not, alas, of cluster analysis, see Sejnowski and Rosenberg 1986.) I include it here for two reasons. The first is that it represents an interesting midway analysis falling between subsymbolic and straightforwardly classical levels of description. The second is that (unlike a mere listing of microfeatures) it is intended to reveal the outlines of the semantic metric embodied in a given network.

NETtalk is a large distributed connectionist model for investigating part of the process of turning written input (words) into phonemic output (sounds or speech). The network architecture consists of a set of input units that are stimulated by seven letters of text at a time, a set of hidden units, and a set of output units that code for phonemes. The output is fed into a voice synthesizer, which produces the actual speech sounds.

The network began with a random distribution of hidden unit weights and connections (within chosen parameters), that is, it had no idea of any rules for converting text to phonemes. Its task was to learn by repeated exposure to training instances to negotiate its way around this particularly tricky cognitive domain (tricky because of irregularities, subregularities, and the sensitivity to context of converting text to phonemes). Learning proceeded in the standard way, i.e., by a back-propagation learning rule. This works by giving the system an input, checking its output (this is done automatically by a computerized "supervisor"), and telling it what output (i.e., what phonemic code) it should have produced. The learning rule then causes the system to minutely adjust the weights on the hidden units in a way that would tend to the correct output. This procedure is repeated many thousands of times. Uncannily, the system slowly and audibly learns to pronounce English text, moving from babble to half-recognizable words and on to a highly creditable final performance.

Cluster Analysis is an attempt to display the shape of the representational space the system has created by the carefully regulated weightings on the hidden unit connections. To see how it works, consider the task of the network to be that of setting hidden unit weights in a way that will enable it to perform a kind of set partitioning. The goal is for the hidden units to respond in distinctive ways when, and only when, the input corresponds to a distinctive output. Thus, in converting text to phonemes, we want the hidden units to perform very differently when given "the" as input than they would if given "sail" as input. But we want them to perform identically if given "sail" and "sale" as inputs. So the task of the hidden units is to partition a space (defined by the number of such units and their possible levels of activation) in a way geared to the job at hand. A very simple system, such as the rock/mine network described in Churchland (forthcoming 1989), may need to partition the space defined by its hidden units into only two major subvolumes, one distinctive pattern for inputs signifying mines and one for those signifying rocks. The complexities of text-to-phoneme conversion being what they are, NETtalk must partition its hidden unit space more subtly (in fact, into a distinctive pattern for each of 79 possible letter to phoneme pairings). Cluster analysis as carried out by Rosenberg and Sejnowski in effect constructs a hierarchy of partitions on top of this base level of 79 distinctive stable patterns of hidden-unit activation. The hierarchy is constructed by taking each of the 79 patterns and pairing it with its closest neighbor, i.e., with the pattern that has most in common with it. These pairings act as the building blocks for the next stage of analysis. In this stage an average activation profile between the members of the original pair is calculated and paired with *its* nearest neighbor drawn from the pool of secondary figures generated by averaging each of the original pairs. The process is repeated until the final pair is generated. This represents the grossest division of the hidden-unit space that the network learned, a division that, in the case of NETtalk, turned out to correspond to the division between vowels and consonants. Cluster analysis thus provides a picture of the shape of the space of the possible hidden-unit activations that power the network's performance.

A few comments. First, it is clear that the clusterings learned by NETtalk (e.g., the vowel and consonant clusterings at the top level) do not involve novel, unheard of subsymbolic features. This may be due in part to the system's reliance on input and output representations that reflect the classical theory. Even so, the metric of similarity built into the final set of weights still offers some clear advantages over a classical implementation. Such advantages will include generalization, various forms of robustness, and graceful degradation.

For our purposes, the most interesting questions concern the status of the cluster-theoretic description. Is it an accurate description of the system's processing? One prominent eliminativist, Churchland (1989), answers firmly in the negative. Cluster analysis, he argues, is just another approximate, high-level description of the system's gross behavior (see my comments on levels 4 and 5) and does not yield an accurate description of its processing. The reason for this is illuminating. It is that the system itself knows nothing about its own clustering profile, and that profile does not figure in the statement of the formal laws that govern its behavior (the activation-evolution and connection-evolution equations of level 1). Thus, Churchland notes, "the learning algorithm that drives the system to new points in weight space does not care about the relatively global partitions that have been made in activation space. All it cares about are the individual *weights* and how they relate to apprehended error. The laws of cognitive evolution, therefore, do not operate primarily at the level of the partitions.... The level of the partitions certainly corresponds more closely to the "conceptual" level..., but the point is that this seems not to be the most important dynamical level" (1989, 25).

Churchland's point is that although a theorist could use her knowledge of a system's clustering profile to predict some of its short-term behavior (e.g., that it will classify inputs *a* and *b* together and *a* and *c* separately), she could not use that knowledge to predict its cognitive development (if it is a learning system) or the precise shape of its possible breakdowns. To get such fine-grained predictive power, Churchland argues, we need to see the precise nature and interrelations of the subconceptual elements responsible for the gross partitioning. By this he does *not* mean that we need the kind of subsymbolic descriptions considered in our discussion of level 2. Rather, he opts for the numerical, connection-weight specification as the appropriate level of analysis.

Level 4, the symbolic-AI level

The term "conceptual level" as used by Smolensky seems to be ambiguous between any construct of classical AI, e.g., a schema, production, prototype, and so on, and the terms of ordinary language, e.g., "table, "office," "coffee," and so on. The two meanings are certainly linked, since some classical AI (the kind I have been calling classical cognitivism) essentially involves the manipulation of entities that have the semantics of words of natural language. However, they seem sufficiently different to merit some teasing apart. Hence, level 4, as I understand it, will be the level at which a connectionist system is described *as if* it were a classical one, e.g., described as firing a production, accessing a schema, extracting a prototype, and so on. (Level 5 will be the level of good, old-fashioned folk-psychological description.)

The general connectionist claim, as we have seen, is that all these conceptual-level constructs of both types offer at best approximately accurate descriptions of the system's behavior. Such constructs (or rather, the explanations in which they figure) give a good indication of what the system will do in a central range of cases, but they are unable to predict or explain various other aspects of the system's capabilities. Recall the connectionist model for solving simple circuit problems detailed in Smolensky 1986 and in chapter 6, section 4 above. The model solves problems by the standard connectionist method of massive parallel satisfaction of a large number of soft constraints, i.e., relations of excitation and inhibition between units that represent subsymbolic features. Nonetheless, it often looks from the outside as if the system must work by satisfying hard, symbolically couched constraints in serial order. If you give it a well-posed problem and unlimited processing time, it will converge on a solution by making a set of micro decisions (recomputations of unit values). And these will in turn contribute to various macro decisions as sections of the network settle into a solution to their part of the problem. These macro decisions appear much like the serial firing of production rules.

But if you give the system an ill-posed problem or artificially curtail its processing time, it still gives what Smolensky calls "sensible performance," This is explained by the underlying subsymbolic nature of its processing, which will always satisfy as many soft constraints as it can, even if given limited time and degraded input. The moral of all this, as Smolensky sees it, is that the theorist may analyze the system at the higher level of, e.g., a set of production rules. This level will capture some facts about its behavior. But in less ideal circumstances the system will also exhibit other behavior that is explicable only by describing it at a lower level. Thus, the *unified* account of cognition lies at one of the lower levels (level 2 or level 1, according to your preference). Hence the famous analogy with Newtonian mechanics. Symbolic AI describes cognitive behavior, much as Newtonian mechanics describes physical behavior. They each offer a useful and accurate account in a circumscribed domain. But the unified account lies elsewhere (in quantum theory in physics, and in connectionism in cognitive science). Thus, commenting on the model for solving circuitry problems, Smolensky notes, "A system that has, at the micro-level, soft constraints satisfied in parallel, *appears* at the macro-level, under the right circumstances to have hard constraints, satisfied serially. But it doesn't *really*, and if you go outside the "Newtonian" domain you see that it's really been a quantum system all along" (1988, 20). Such "Newtonian" analyses are conceded to be useful in that they may help describe interrelations between complex patterns of activity that approximate various conceptual constructs in which the theorist is interested. As Smolensky points out (1988, 6), such interactions will not be "directly described by the formal definition of a subsymbolic model"; instead, they must be "computed by the analyst."

Level 5, the folk-psychological level

The folk-psychological level is the level in which we use the words and concepts of ordinary language in the ordinary way to describe the cognitive states of a system. Thus, just as we may say, "John believes that Mary will get the chair," so we may say, "The network believes that bedrooms contain dressing tables." When we are dealing with a toy network, it is clear enough that the system is in some way not a proper object of full-blooded belief ascription. But in considering the status of folk-psychological description, we must bracket this fact and ask instead the following question. If, as Smolensky says (1988, 7), connectionism does indeed afford the "complete formal account of cognition," what follows on the status of folk-psychological descriptions of human mental states?

Drawing on our previous discussion, we can immediately observe that if the human mind is a pure distributed connectionist system, then the individual words used in a belief ascription will not have discrete, recurrent analogues in the actual processing of the system. Thus, the word "chair" will not have a discrete analogue, since "chair" will be represented as an activation vector across a set of units that stand for subsymbolic microfeatures, and it will not have a single recurrent analogue (not even as an activation vector), since the units that participate and the degree to which they participate will vary from context to context.

The radical eliminativist takes these facts and conjoins them with a condition of causal efficacy, which states: a psychological ascription is only warranted if the items it posits have direct analogues in the production (or possible production) of behavior. Thus ascribing the belief that cows can't fly to John is justified only if there is some state in John in which we can in principle identify a discrete, interpretable substate with the meaning of "cow," "fly," and so on. Since, according to connectionism, there are no such discrete, recurrent substates, the radical eliminativist concludes that commonsense psychology is mistaken and does not afford an accurate higher-level description of the system in question (John). This is *not* to say that such descriptions are dispensable in practice; it is to say only that they are mistaken in principle.

In the next section I shall sketch an account of explanation that dissociates the power and accuracy of higher-level descriptions from the condition of causal efficacy, which thereby gives a more liberal, more plausible, and more useful picture of explanation in cognitive science and daily life.

3 *Explanation Revisited*

The eliminativist argues her case as follows.

> *Step 1.* Suppose that pure distributed connectionism offers a correct account of cognition.
>
> *Step 2.* It follows that there will be no discrete, recurrent, in-the-head analogues to the conceptual-level terms that figure in folk-psychological belief ascription.
>
> *Step 3.* Hence by the condition of causal efficacy, such ascriptions are not warranted, since they have no in-the-head counterpart in the causal chains leading to action.
>
> *Step 4.* Hence, the causal explanations given in ordinary terms of beliefs and desires (e.g., "She went out because she believed it was snowing") are technically mistaken.

My claim will be that even if pure distributed connectionism offers a correct and (in a way) complete account of cognition, the eliminativist conclusion (step 4) doesn't follow. It doesn't follow for the simple reason that good causal explanation in psychology is *not* subject to the condition of causal efficacy. Likewise, even if pure distributed connectionism is true, it does not follow that the stories told by symbolic AI are mere approximations. Instead, I shall argue, these various vocabularies (e.g., of folk-psychology and of symbolic AI) are geared accurately to capture legitimate and psychologically interesting equivalence classes, which would be *invisible* if we restricted ourselves to subsymbolic levels of description. In a sense, then, I shall be offering a version of Dennett's well-known position on folk-psychological explanation but extending it, in what seems to me to be a very natural way, to include the constructs of symbolic AI (e.g., schemata, productions.) If I am right, it will follow that many defenders of symbolic AI and folk psychology (especially Fodor and Pylyshyn) are effectively shooting themselves in the feet. For the defences they attempt make the condition of causal efficacy pivotal, and they try to argue for neat, in-the-head correlates to symbolic descriptions (see, e.g., Fodor 1987; Fodor and Pylyshyn 1988). This is accepting terms of engagement that surely favor the eliminativist and that, as we shall see, make nonsense of a vast number of perfectly legitimate explanatory constructs.

What we need, then, is a notion of causal *explanation* without causal *efficacy*. I tried for such a notion in Clark, forthcoming. But a superior case has since been made by Frank Jackson and Philip Pettit, so I begin by drawing on their account. Jackson and Pettit ask the reader to consider the following case. "Electrons A and B are acted on by independent forces F_A and F_B respectively, and electron A then accelerates at the same rate as electron B. The explanation of this fact is that the magnitude of the two forces is the same.... But this sameness in magnitude is quite invisible to A.... This sameness does not make A move off more or less briskly" (1988, 392–393). Or again, "We may explain the conductor's annoyance at a concert by the fact that someone coughed. What will have actually caused the conductor's annoyance will be the coughing of some particular person, Fred, say" (Jackson and Pettit 1988, 394). This is a nice case. For suppose someone, in the interests of "accuracy," insisted that the proper (fully causal) explanation of the conductor's annoyance was in fact *Fred's* coughing. There is a good sense in which their "more accurate" explanation would in fact be *less powerful.* For the explanation which uses "someone" has the advantage of making it clear that "any of a whole range of members of the audience coughing would have caused annoyance in the conductor" (Jackson and Pettit 1988, 395). This increase in generality, bought at the cost of sacrificing the citation of the actual entity implicated in the particular causal chain in question, constitutes (I want to say) an explanatory virtue, and it legitimizes a whole range of causal explanations that fail to meet the condition of causal efficacy. Likewise in the electron case, there is no analogue of "sameness" which propels electron *A*. But citing sameness in our causal explanation highlights the fact that the same result (identical acceleration) would be obtained by an infinite number of values of F_A and F_B provided just that F_A is equal to F_B.

A final example, especially relevant for our subsequent discussion, is Hilary Putnam's peg and hole explanation. We explain the fact that a one inch square peg won't pass through a one inch round hole by citing the squareness of the peg and the general fact that squareness will not pass through an equivalent area of roundness. Yet suppose we ask for the "real" causal story. In any given case this will involve a mass of subatomic facts about clouds of particles. But exclusively to focus on this is to obscure the whole range of situations that we might well be interested in grouping together as cases in which squareness won't pass through roundness. (Imagine, for example, a universe with a different microstructure, but one which still sustains higher-level descriptions in terms of roundness and squareness.) The moral is that "in no particular case will the squareness and the roundness as such figure in the full story of the multitude of interactions which stop the peg from fitting into the hole, but the fact of squareness and roundness ensures, though not causally, that there is some very complex set of interactions which stops the peg from fitting into the hole" (Jackson and Pettit 1988, 395).

I hope that this talk of multitudes of smaller interactions and causally inactive higher-level descriptions puts the reader in mind of the subsymbolic and higher-level descriptions of connectionist systems laid out in section 2. Such, at any rate, will be the line I play out in the next section. First, though, a little borrowed terminology. After Jackson and Pettit, let us call an explanation that highlights a common feature of a range of cases (e.g., the explanations that cite roundness and sameness) but abstracts away from the causally active features of a particular case a *program explanation.* In these styles of explanation the common feature or property will be said to causally program the result without actually figuring in the causal chain leading to an individual action or instance. And let us contrast such explanations with *process explanations,* which cite the very features that are efficacious in a particular case or range of cases. My claim, then, is that explanations using the various higher-level constructs of symbolic AI and folk psychology may be necessary and fully accurate program explanations, while failing (as the eliminativist insists) to constitute good process explanations. They will do this just in case they offer a terminology that groups various systems into psychologically interesting equivalence classes that are unmotivatable if we restrict ourselves to, say, a pure subsymbolic account of processing.

4 *The Value of High-Level Descriptions*

Consider once again the various higher-level analyses of pure distributed connectionist systems. The first of these was cluster analysis. Cluster analysis, recall, involved charting the hierarchy of divisions (or partitions) that the network had learned to make using its hidden units. Recall also the attitude of at least one leading eliminativist, Paul Churchland, to such a level of analysis. It was that when undergoing conceptual change (e.g., learning) the system would behave in ways not responsive to the various partitionings (which the system does not really know about), but it would behave in ways responsive to the actual connection weights (which it does really know about).

This point is well taken as far as it goes. (It is like saying, "If you want to predict the *actual* acceleration of electron *A*, you'd better know the values of the forces acting on it and not just that they are the same as those acting on electron *B*.") But it would be a grave mistake to assume that this point shows that the level of analysis adopted by the cluster analyst is inferior, approximate, unnecessary, or downright mistaken. For an analysis that cites partitionings, like one that cites the sameness of the forces acting on the electrons, may likewise have virtues that other analyses cannot reach. For example, it is an important fact about cluster analysis (a fact recognized by Churchland [1989, 24]) that networks that have come to embody different connection weights may have identical cluster analyses. Thus, Sejnowski notes that versions of NETtalk that begin with different random distributions of weightings on the hidden units will, after training, make the same partitions but by means of different arrangements of weights on the individual connections. Now consider a particular cognitive domain, say, converting text to phonemes. Isn't it a legitimate psychological fact that only certain systems can successfully negotiate that domain? And don't we want some level of properly psychological, or cognitive, explanation with the means to group such systems together and to make some generalizations about them (e.g., that such systems will be prone to certain illusions)? Cluster analysis is the very tool we need, it seems. Any one of a whole range of networks, we can say, will be able to negotiate that cognitive domain. And we can give an account that specifies what networks belong in that range (or in the equivalence class in question) by requiring that they have a certain cluster analysis. In the terminology introduced in section 3, the cluster analysis causally programs the system's successful performance, but it is not part of any process explanation.

Let us now move up another level to the descriptions offered by symbolic AI. Suppose, for the sake of argument, that we describe NETtalk at this level as a discrimination tree, or better, as a production system with one production for each conversion of text to phoneme that it has learned. We have clearly lost explanatory power for explaining the performance of an individual network. For as we saw, the network will perform well with degraded information in a way that cannot be explained by casting it as a standard symbolic AI system. But as with the cluster analysis, we gain something else. For we can now define an even wider equivalence class that is still, I suggest, of genuine psychological interest. Membership of this new, wider class requires only that the system behave in the ways in which the pure production system would behave in some central class of cases. The production-system model would thus act as an anchor, dictating membership of an equivalence class, just as the cluster analysis did in the previous example. And the benefits would be the same too. Suppose there turns out to be a lot of systems (some connectionist, some classical, some of kinds still undreamed of) all of which nonaccidentally approximate the behavior of the pure production system in a given range of cases. They are all united by being able to convert text to phonemes. If we seek some principled and informative way of grouping them together (i.e., not a bare disjunction of systems capable of doing such and such), we may have no choice but to appeal to their shared capacity to approximate the behavior of such and such a paradigmatic system. We can then plot how each system manages, in its different way, to approximate each separate production. Likewise, there may be a variety of systems (some connectionist, some not) capable of supporting knowledge of prototypical situations. The symbolic AI construct of a schema or frame may help us understand in detail, beyond the gross behavior, what all these systems have in common (e.g., some kind of content addressability, default assignment, override capacity, and so forth). In short, we may view the constructs of symbolic AI, not as mere approximations to the connectionist cognitive truth, but as a means of highlighting a higher level of unity between otherwise disparite groups of cognitive systems. Thus, the fact that a connectionist system *a* and some architecturally novel system of the future *b* are both able to do commonsense reasoning may be explained by saying that the fact that *a* and *b* each approximate a classical script or frame-based system causally programs their capacity to do commonsense reasoning. And this means that a legitimate higher-level shared property of *a* and *b* is invisible at the level of a subsymbolic analysis of *a*. This is not to say, of course, that the subsymbolic analysis is misguided. Rather, it is to claim that that analysis, though necessary for many purposes, does not render higher levels of analysis defunct or of only heuristic value.

Finally, let us move on to full-fledged folk-psychological talk. On the present analysis, such talk emerges as just one more layer in rings of evermore explanatory virtue. The position is beautifully illustrated by Daniel Dennett's left-handers thought experiment. "Suppose," Dennett says, "that the sub-personal cognitive psychology of some people turns out to be dramatically different from that of others." For example, two people may have very different sets of connection weights mediating their conversions of text to phonemes. More radically still, it could be that left-handed people have one kind of cognitive architecture and right-handed people another. For all that, Dennett points out, we would never conclude on those grounds alone that left-handers, say, are incapable of believing.

> Let left- and right-handers be as internally different as you like, we already know that there are reliable, robust patterns in which all behaviourally normal people participate—the patterns we traditionally describe in terms of belief and desire and the other terms of folk psychology. What spread around the world on July 20th, 1969? The belief that a man had stepped on the moon. In no two people was the effect of the receipt of that information the same ..., but the claim that therefore they all had nothing in common ... is false, and obviously so. There are indefinitely many ways one could reliably distinguish those with the belief from those without it. (Dennett 1987, 235)

In other words, even if there is no single internal state (say, a sentence in the language of thought) common to all those who are said to believe that so and so, it does not follow that belief is an explanatorily empty construct. The sameness of the forces acting on the two electrons is itself causally inefficacious but nonetheless figures in a useful and irreducible mode of explanation (program explanation), which highlights facts about the range of actual forces that can produce a certain result (identical acceleration). Just so the posit of the shared belief highlights facts about a range of internal cognitive constitutions that have some common implications at the level of gross behavior. This grouping of apparently disparate physical mechanisms into classes that reflect our particular interests is at the very heart of the scientific endeavor. To suppose that the terms and constructs proper to such program explanations are somehow inferior or dispensable is to embrace a picture of science as an endless and disjoint investigation of individual causal mechanisms.

There is, of course, a genuine question about what constructs best serve our needs. The belief construct must earn its keep by grouping together creatures whose gross behaviors really do have something important in common (e.g., all those likely to harm me because they *believe* I am a predator). In recognizing the value and status of program explanations I am emphatically *not* allowing that anything goes. My goal is simply to counter the unrealistic and counterproductive austerity of a model of explanation that limits "real explanations" to those that cite causally efficacious features. The eliminativist argument, it seems, depends crucially on a kind of austerity that the explanatory economy can ill afford.

5 *Self-Monitoring Connectionist Systems*

The previous four sections have, I hope, established that even if pure distributed connectionism constitutes a complete and accurate formal model of cognition (as Smolensky claims), it does not follow that higher-level analyses (like cluster analysis, symbolic AI, and folk psychology) are misguided, mistaken, or even mere approximations. Instead, they may be accurate and powerful *grouping* explanations of the kind examined in sections 3 and 4.

In this more speculative section I want to use the same kind of observations to cast some doubt on the idea that pure distributed connectionism constitutes a complete and accurate formal model of cognition. There is a very natural development here, since the virtues of grouping explanations as third-person, theorist's constructs have analogues in first-person processing. In short, there may be pressure on individual cognizers to monitor and group their own internal states in the same general way as there is pressure for program explanations that group other people's internal states. If this is the case, then parts of our internal cognitive economy may begin to look distinctly classical and symbolic. Without presuming to decide what is clearly an empirical issue about cognitive architecture, this section aims to depict the kinds of pressure that might make such a mixed cognitive economy attractive.

Pure distributed connectionism insists that discrete symbolic constructs (e.g., "dog," "office") exist *only* as instruments of interpersonal communication in a public language and as constructs of higher-level, third-person analyses of shifting, fluid activation vectors. Let us bracket for now the question of public language. All theorists agree that we use and process at some level the symbolic entities of public discourse (words). My question is whether such symbolic entities (discrete recurrent items and conceptual semantics) have any role to play in individual cognition beyond whatever is necessary to produce and interpret language. The pure distributed connectionist thinks not. She agrees, with Smolensky, that such entities at best emerge at a higher level of analysis of what are through and through subsymbolic systems (see, e.g., Smolensky (1988, 17). That is to say, such "entities" are visible only to the external theorist and do not figure in the system's own inner workings. All the system knows about, on this account, are its manifold activation vectors. The rest are theorist's fictions, a useful and (according to our earlier arguments) even indispensable aid to grouping systems into equivalence classes, but not a feature of individual processing (recall Churchland's comment about cluster analysis revealing partitions that the system itself knows nothing about).

Pure distributed connectionism, we saw in chapter 9, is heir to some interesting problems. For it seems that systems lacking the distinctive resources of classical symbolic AI (hard pattern matching, variable binding, easy recombination of atomic elements, etc.) may be inadequate for some tasks. In their less excessive moments, such recent criticisms as Fodor and Pylyshyn 1988 and Pinker and Prince 1988 may be touching on just such inadequacies. In particular, there may be a set of problems associated with the apparent lack, in pure distributed connectionism, of anything corresponding to higher-level *labels* for sets of distributed activity, labels that are capable of acting either as cues for the full distributed representation or as stand-ins in operations (e.g., deductive inference) in which the full distributed representation is either unnecessarily complex or even distorting because of its extreme context sensitivity (see criticisms in Fodor and Pylyshyn 1988). I propose to sketch three problem areas that are representative of this kind of pressure for actual in-the-head symbolic structures.

The first is what I shall call the problem of relying on cues. This problem (which is the subject of detailed investigation in Robbins, unpublished) can be introduced with a simple example. Consider a pure distributed connectionist model of prototype extraction (e.g., McClelland and Rumelhart 1986). Such a network will be given many examples of a certain kind of item. For instance, it may be given many examples of dogs, each particular dog being specified as a set of microfeatures. Since connectionist models must store data superpositionally by the subtle orchestration of a whole set of units each of which participates in the encoding of many patterns, the network is able to generate a prototypical dog representation, as we saw in chapter 5. This representation is just the center of the state space defined by the set of dog inputs. Such inputs will have various features in common, and these will become strongly represented in the network's pattern-completion dispositions. The more variable features will tend to cancel out, unless recalled by a unique cue. A further benefit is that one set of weights can encode multiple prototypes (e.g., of a dog, a cat, and a bagel) if they are sufficiently dissimilar to avoid confusion (see McClelland and Rumelhart 1986, 185).

The trouble with all this, as Robbins points out, is that as things stand the system "knows" the prototype in only a very weak sense. In knows it in that given, say, half the prototypical dog as a cue, it could reliably complete the rest. But it has no way of representing the prototype to itself *as* a prototype. In a sense, the center of its own state space is as invisible to the system itself as are the higher-level partitionings mentioned earlier. The system "knows" about prototypical dogs in the way a musical novice with a good ear may "know" about E-flat. The novice can differentially respond to E-flat inputs, but she has not marked the sound out to herself.

One response to this, suggested by McClelland and Rumelhart, is to associate a name (e.g., "dog") with all the exemplar inputs. The result would be that "dog" would then act as a partial cue capable of generating a full prototypical representation. This move, however, simply defeats the purpose of PDP, for the network now does not perform anything like automatic categorization. Instead the theorist, in training the network, decides what categories it will really know about by feeding it the labels.

What is actually required, I suggest, is some kind of internal pressure built into the system so that it is driven to seek particularly salient peaks of activation and to label them itself as part of a process of monitoring and organizing itself. Salience, as Robbins also points out, is not the same as any old signal average or any old state space center. Instead, we want the system itself to recognize the center of some state space as a particularly important activation vector and then label it for future use.

There is an analogy here with some of Pinker and Prince's (1988) criticisms of the past-tense-learning network reviewed in chapter 9. That network was forced into a higher rule-based level of organization by *external* pressure (in fact, by the sudden influx of a large body of regular past tenses). In contrast, Pinker and Prince argue, the child is *internally* driven to seek principles by which to organize and regiment the examples she already knows. This posit of internal pressure for a real higher-level understanding fits very well with some speculations of Annette Karmiloff-Smith, a leading developmental psychologist. The child, according to Karmiloff-Smith (1985, 1986, 1987), is her own metatheorist. The child tries to bring her own processing strategies under higher-level descriptions in order to facilitate problem solving. This effect amounts to doing for one's own processing what program explanations were seen to do for the processing of groups of systems; it amounts to seeking important common effects of various strategies and then explicitly grouping such strategies into labeled equivalence classes. One area in which the availability to oneself of such higher-level analyses of one's own processing is especially useful is in debugging one's own performances.

This leads very nicely to the second kind of problem I want to mention: what Smolensky calls the "assignment-of-blame problem." The assignment-of-blame problem is that of deciding, when a system fails to perform in some desired way, what features of the system are at fault. As Smolensky notes, "In subsymbolic systems, this assignment of blame problem is a difficult one, and it makes programming subsymbolic models by hand very tricky" (1988, 15). Automatic learning procedures (like back-propagation) constitute a partial solution to the problem. But if, after a period of training, something is still going wrong, it is extremely hard to put it right except by more training on better chosen examples. In short, you can't really debug a pure distributed connectionist network.

Compare this state of affairs to the position of the human expert. A golfer say, who has a problem with her swing. The golfer will naturally wish to debug her swing in the most efficient way possible. And this will involve not just practice but very carefully chosen practice, practice aimed at some aspect of her swing that she feels is the root of the trouble (say, wrist control). For the expert golfer, having a higher-level articulation of the swing into wrist, arm, and leg components is an essential aid to improving and debugging performance.

What's true of golf is often true of life in general, and the present case is no exception. The really expert cognizer, I suggest, builds for itself various higher-level representations of its own lower-level (pure distributed connectionist) reasoning. These representations (which must group activation vectors into rough classes according to their various roles in producing behavior) provide the key to efficient improvement and debugging. The studies by Karmiloff-Smith referred to above provide evidence that just such a process of higher-level representation occurs in children. Thus, she shows how various aspects of linguistic competence seem to arise in three phases. The first phase involves the child's attainment of basic behavioral success. The child learns to produce the required linguistic forms. Internally driven by a goal of control over the organization of internal representations, the child then goes on to establish a structured description of her own basic processing (phase two). "The initial operation of phase two is to re-describe the phase one representations in a form which allows for (albeit totally unconscious) access" (Karmiloff-Smith 1986, 107). This added cognitive load, however, may cause new errors, which are only corrected once a balance of phase 1 (procedural competence) and phase 2 (redescriptive competence) is achieved. This balancing act constitutes phase 3. The general message (backed by a wealth of data in Karmiloff-Smith 1985, 1986, 1987) is that the child (and also the adult, for this is a general learning pattern) first learns to produce correct outputs, and then yields to endogenous pressure to form representations of the form of the processing that yields the outputs. These higher-level representations are much closer to classical discrete symbol structures than to highly distributed connectionist representations.

Perhaps, then, the pure distributed connectionist underrates the value of the discrete symbol structures used in natural language and classical AI. Such structures, according to the radical connectionist, serve at best two roles, to mediate interpersonal communication and to help the novice acquire rudimentary skills in a domain. I have argued, in contrast, that such symbol structures may also be a vital aid to the *expert* by providing an articulated model of her own problem-solving strategies. This model, though lacking the fluidity and power provided by the connectionist substratum, may be an indispensable aid to debugging one's own performance.

Finally, I wish to consider an internal correlate to the role of interpersonal communication. Just as the gross symbol structures of public language facilitate communication between whole agents who may have very different internal representations of the states of affairs being discussed, so internal symbol structures may facilitate communication between different connectionist subsystems that have very different representations of some domain of mutual interest.

It seems plausible to suppose that the mind is not one big undifferentiated network. Instead we should expect some articulation into special-purpose networks. This immediately brings a problem of communication. It would be intolerably wasteful, for example, for a network that needs to coordinate its activity with some other network to have to reproduce in full the activation vectors characteristic of the other network's processing. It is likely to need only a sketch of the overall shape of the other's activity. This could easily constitute pressure for a kind of higher-level representational formalism geared to pass messages between the various subsystems of a single cognizer. Thus, Churchland imagines that "two distinct networks whose principal concerns and activities are non-linguistic learn from scratch some systematic means of manipulating, through a proprietory dimension of input, the cognitive abilities of the other network." He then asks, "What system of mutual manipulation—what language—might they develop?" (Churchland 1989, 42).

In sum, then, there seems to be significant *internal* pressure to develop various higher-level articulations of subsymbolic activation vectors. Instead of being just an external theorist's description of emergent features of subsymbolic processing, many symbolic constructs may be incarnate within the system as its own description of emergent features of its subsymbolic processing. Such a conjecture seems to me to go some way toward fleshing out some of Daniel Dennett's recent claims about the evolution of consciousness. This is obviously a massive topic, which I can barely touch on here. But part of Dennett's picture is of a form of consciousness in which some animals simulate a Von Neumann, discrete, serial processor using their natural subsymbolic connectionist cognitive architecture. The usefulness of such a development, Dennett thinks, may be to create a system that can benefit from its own self-monitoring activity. Dennett's story involves some elaborate, though to my mind rather plausible, conjectures on the role of self-stimulation in the creation of such an architecture. I shall not attempt to detail his argument here. The upshot of it is "an architecture that is incessantly re-organising itself, trying out novel combinations, sometimes idly, sometimes with great purpose.... And what is all this good for? It seems to be good for the sorts of self-monitoring that can protect a flawed system from being victimised by its own failures" (Dennett 1988, 27). This insistence on the importance of self-monitoring and on its association with a somewhat more classical style of architecture and representation chimes very nicely with my picture of self-improving, debugging systems that utilize condensed or higher-level representations to engage in metareasoning about their own basic processing.

Of course, many connectionists have already felt the need to introduce coarser representations than those characteristic of a pure distributed approach. We saw that Geoffrey Hinton, for example, has suggested that it may be necessary to equip systems with two representations of everything. One would be the full, distributed representation spread across a whole set of feature units. The other would be a condensed version capable of standing in for and recalling, if necessary, the full (or expanded) representation. This, like Minsky's k-line theory (1980) and various other partitially local, partially distributed approaches, can be seen as a way of responding to the pressures sketched above. If these pressures are genuine, and if the answer is some kind of mixed system (part distributed, part symbolic), then it seems that the raging debate over the "correct" cognitive architecture is badly posed. For if human beings are indeed mixed systems employing various kinds of representations and operations for various purposes, they will be properly described in both classical and connectionist terms. The serious questions would then arise case by case for every aspect of cognition, and the absolutely global question (PDP or not PDP?) would wither and die, a vestigial organ proper to a previous generation of debate.

6 *A Double Error*

The eliminativist, I have argued, makes a double error. First, her conditional argument is flawed. Even if pure distributed connectionism *were* a complete, formal account of individual processing, it would not follow that the higher-level constructs of symbolic AI and folk psychology were inaccurate, misguided, or dispensable. Instead, such constructs may be the essential and accurate grouping principles for explanations which prescind from causal process to causal program. The eliminativist's conditional argument was seen to rest on an insupportable condition of causal efficacy, a condition that, if applied, would rob us of a whole range of perfectly intelligible and legitimate explanations both in cognitive science and in daily life.

Second, the antecedent of the eliminativist's conditional is itself called into doubt by the power and usefulness of higher-level articulations of processing. Such articulations, I argued, may do in the first person what program explanations do in the third. That is, they may enable the system to observe and mark the common contributions of a variety of activation vectors. Such marking would provide the system with the resources to reflect on (and hence improve and debug) its own basic processing strategies. In this way, entities that the pure distributed connectionist views as emergent at a higher level of *external* analysis (high-level clusterings, types, and categories) may in fact be *incarnate* in virtue of the system's own self-analytic activity.

If this is correct, a notable implication is that the whole "What architecture?" debate turns out to have been seriously ill posed. Where the question once seemed to be PDP or not PDP? it now becomes merely, Where PDP, and where not PDP? Thus may the ungrammatical triumph over the unanswerable.

Notes

Chapter 1

1. On the pro side, see Pylyshyn 1986. On the con side, Searle 1980 and Dreyfus 1981, both discussed at length in chapter 2. For rare philosophical treatments sensitive to alternative computational approaches, see Boden 1984b; Davies, forthcoming; Hofstadter 1985; and, somewhat surprisingly, Dreyfus and Dreyfus 1986.
2. See, for example, Marr and Poggio 1976; Hinton and Anderson 1981; and McClelland, Rumelhart, and the PDP Research Group 1986, as well as almost any work in computational linguistics.
3. For thorough summaries see Haugeland 1985, Hodges 1983, Turing 1937, and Turing 1950.
4. In fact, the idea of list processing (in which data structures contain symbols that point to other data structures, which likewise contain symbols that point to other data structures and so on, thus facilitating the easy association of information with symbols) was introduced by Allan Newell and Herbert Simon in their program Logic Theorist of 1956. For a detailed treatment of list processing see E. Charniak and D. McDermott 1985, chapter 2, and also A. Newell and H. Simon 1976. One attractive and important feature of LISP was its universal function *eval*, which made it as adaptable as a universal Turing machine, barring constraints of actual memory space in any implementation.
5. A production system is essentially a set of "if-then" pairs in which the "if" specifies a condition that, if satisfied, causes an action (the "then") to be performed. Each "if-then" pair or condition-action rule is called a production (see, for example, Charniak and McDermott 1985, 438–39).
6. The AM program, as Lenat himself later pointed out, worked partly due to the amount of mathematical knowledge implicit in LISP, the language in which it was written (see Lenat 1983a; Lenat 1983b; Lenat and Brown 1984; Ritchie and Hanna 1984). This result seems related to some observations I make later concerning the amount of the work of scientific discovery already done by giving BACON data arranged in our representational notation. Lenat's later work, EURISKO, avoids the "defect" of trading on the quasi-mathematical nature of LISP. But it too relies on our giving it notationally predigested data. I do not see this as a defect: human babies inherit a rather impressive notation in the form of a well-structured public language. I still affirm the comments in chapter 10 that we need the right substructure below the public notation when instantiation, not just psychological explanation, is the issue.
7. I am especially indebted to Martin Davies for many fruitful conversations on the topic.

Chapter 2

1. Thanks to Lesley Benjamin for spotting the shampoo example and for some stimulating conversation about the problems involved in parsing it.

2. See, e.g., Torrance 1984, 23. In fairness to Dreyfus, he goes on to say that his point is not that humans can recognize subtleties currently beyond the reach of simple programs but that "in any area there are simple taken-for-granted responses central to human understanding, lacking which a computer program cannot be said to have any understanding at all" (Dreyfus 1981, 190). As will become apparent, I agree with this but believe the real point concerns a degree of *flexibility* of behavior, which cannot be modeled by the SPSS approach. This concerns not exactly *what* is and is not known but rather *how* it is known. And this fits better with the second line of Dreyfus's argument than the first.
3. These replies are well known (see Searle 1980 and, e.g., Torrance 1984). I do not think that the discussion here has been very illuminating, and I therefore choose to ignore it for present purposes.
4. I am indebted to Aaron Sloman for teaching me to put this point in terms of structural variability.
5. This was pointed out to me by Michael Morris in conversation.

Chapter 3

1. At a minimum, the eliminative materialist must believe that this is the *primary* point of the practice. In conversation Paul Churchland has accepted that folk-psychological talk serves a variety of other purposes also, e.g., to praise, to blame, to encourage, and so on. But, he rightly says, so did witch talk. That in itself is not sufficient to save the integrity of witch talk. I agree. The primary point of witch talk, however, was to pick out witches. In this it failed, since there were none (we believe). Comparably, Churchland holds that the primary purpose of folk-psychological talk is to fix on neurophysiologically sound states of the head so as to facilitate the prediction and explanation of others' behavior or (more properly) their bodily movements. I deny that this is the primary purpose.
2. I am especially grateful for the simple but suggestive picture of mental-state ascriptions that exploit the cheap and available resource of talk about the external world.
3. As used by (among others) Aaron Sloman, Barry Smith, and Steve Torrance.

Chapter 5

1. Aaron Sloman has fought hard to convince me of this. I thank him for his persistent caution.
2. See especially pages 27–31.
3. This example was originally developed in McClelland 1981 and subsequently redeployed in McClelland, Rumelhart, and the PDP Research Group 1986. My presentation is based on the 1986 version.
4. Philosophically, this talk of representation is lax and misleading. I, like many others, use it because it is useful and concise. What I *mean* by representation here and in similar occurrences elsewhere is just the part of a system's internal structure that is particularly implicated in an environmentally embedded system's capacity to behave in ways that warrant talk of that system's having a representation of whatever is in question. This is quite a mouthful, as you can see. And the notion of what is "particularly implicated" in the relevant behavior is intuitively clear but hard to make precise.
5. For a full account of how such learning rules work (e.g., the generalized delta rule and the Boltzmann learning rule), see McClelland, Rumelhart, and the PDP Research Group 1986, vol. 1, chapters 7 and 8.

Chapter 6

1. A related issue here concerns our capacity to change verbs into nouns and vice versa. Thus newly coined uses like "She wants to thatcher the organization" or "Don't crack wise with me" are easily understood. This might be partially explained by supposing that the verb/noun distinction is (at best) one microfeature among many and that other factors, like position in the sentence, can force a change in this feature assignment while leaving much of the semantics intact. (This phenomenon is dealt with at length in Benjamin, forthcoming).
2. Many of the ideas in this section (including the locution "an equivalence-class of algorithms") are developed out of conversations with Barry Smith. He is not to blame, of course, for the particular views I advance.

Chapter 7

1. See, e.g., Hinton 1984.
2. Lecture given to the British Psychological Society, April 1987.
3. This chapter owes much to Hofstadter (1985), whose suggestive comments have doubtless shaped my thought in more ways than I am aware. The main difference, I suspect, is that I am kinder to the classical symbolic accounts.

Chapter 8

1. As ever, this kind of claim must be read carefully to avoid Church-Turing objections. A task will count as beyond the explanatory reach of a PDP model just in case its performance requires that to carry it out, the PDP system must simulate a different kind of processing (e.g., that of a Von Neumann machine).
2. He allows that intentional realism *can* be upheld without accepting the LOT story (see Fodor 1987, 137). But as I suggest in section 3, he *does* seem to believe that *some form* of *physical* causation is necessary for the truth of intentional realism. This, I argue, is a very dangerous assumption to make.
3. A parsing tree is a data structure that splits a sentence up into its parts and associates those parts with grammatical categories. For example, a parsing tree would separate the sentence "Fodor exists" into two components "Fodor" and "exists" and associate a grammatical label with each. These labels can become highly complex. But the standard simple illustration is

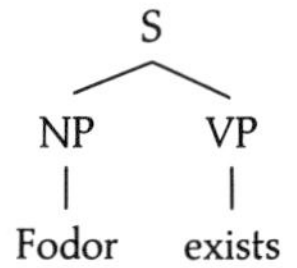

in which S = sentence, NP = noun phrase, and VP = verb phrase.
4. Actually, the case is somewhat complicated by the occasional need to reject the competent speaker's judgments about grammaticality. This occurs when a sentence is rejected due to semantic oddness, length, or complexity. The descriptivist is thus committed to generating the sentences that a certain kind of *ideal* speaker would judge to be grammatical. But these caveats can be entered without undermining the descriptivist's avowed

lack of interest in actual brain processing mechanisms by adverting to grounds of broad theoretic simplicity (see Stich 1972, 211).

5. An exception may be the case of *conscious* thought. We may consciously deploy, say, a mental model of fluid flow to understand electricity or consciously apply grammatical rules in the early stages of speaking a second language or consciously reason by adducing a series of sententially formulated facts and rules. In such cases, it seems natural to assume that at least some level of computational organization in the brain *does* involve operations on formal tokens having a projectible semantics that can be given using the concepts and relations appearing in sentential formulations. As for the other cases, which form the bulk of daily life, we can only wait and see what develops.

Chapter 9

1. The actual structure of the model is complicated in various ways not germane to present concerns. See Rumelhart and McClelland 1986 for a full account.
2. A trivial model would be one that merely used a PDP substrate to implement a conventional theory. But there are complications here; see section 5.
3. This example is mentioned in Davies, forthcoming, 19.
4. Thanks to Martin Davies for suggestive conversations concerning these issues.
5. I owe this suggestion to Jim Hunter.
6. This point was made in conversation by C. Peacocke.
7. For example, Sacks reports the case of Dr. P., a music teacher who, having lost the holistic ability to recognize faces, makes do by recognizing distinctive facial features and using these to identify individuals. Sacks comments that the processing these patients have intact is machinelike, by which he means like a conventional computer model. As he puts it:

   > Classical neurology ... has always been mechanical.... Of course, the brain is a machine and a computer..., but our mental processes which constitute our being and our life, are not just abstract and mechanical [they] involve not just classifying and categorising, but continual judging and feeling also. If this is missing, we become computer-like, as Dr. P. was.... By a sort of comic and awful analogy, our current cognitive neurology and psychology resembles nothing so much as poor Dr. P.! (Sacks 1986, 18–19)

   Sacks admonishes cognitive science for being "too abstract and computational." But he might as well have said "too rigid, rule-bound, coarse-grained, and serial."

Chapter 10

1. This is not to say that the philosophers who raised the worries will agree that they are best localized in the way I go on to suggest. They won't.

Epilogue

1. The story is inspired by two sources: the Gould and Lewontin critique of adaptationist thinking, reported in chapter 4, and Douglas Hofstadter's brief comments on operating systems (1985, 641–642).

Bibliography

Adams, D. 1985. *So long and Thanks for All the Fish.* London: Pan.

Andler, D. 1988. Representations in cognitive science: Beyond the Pro and the Con. *CREA* research paper, Paris.

Armstrong, D. 1970. The nature of mind. Reprinted in N. Block, ed., *Readings in Philosophy of Psychology*, vol. 1, pp. 191–199. London: Methuen and Co., 1980.

Baddeley, R. 1987. Connectionism and gestalt theory. Unpublished manuscript. University of Sussex.

Baron-Cohen, S., Leslie, A., and Frith, U. 1985. Does the autistic child have a "theory of mind"? *Cognition* 21: 37–46.

Benjamin, L. Unpublished. How nouns can verb. Draft research paper. University of Sussex.

Block, N. 1980. Troubles with functionalism. In N. Block, ed., *Readings in Philosophy of Psychology*, vol. 1, pp. 268–305. London: Methuen and Co.

Bobrow, D., and Winograd, T. 1977. An overview of KRL, a knowledge representation Language. *Cognitive Science* 1: 3–46.

Boden, M. 1984a. Animal perception from an AI viewpoint. In C. Hookway, ed., *Minds, Machines and Evolution.* Cambridge: Cambridge University Press.

Boden, M. 1984b. What is computational psychology? *Proceedings of the Aristotelian Society*, suppl. 58: 17–35.

Brady, M., Hollenbach, J., Johnson, T., Lozano-Perez, T., and Mason, M., eds. 1983. *Robot Motion: Planning and Control.* Cambridge: MIT Press.

Broadbent, D. 1985. A question of levels: Comment on McClelland and Rummelhart. *Journal of Experimental Psychology: General* 114: 189–192.

Charniak, E., and McDermott, D. 1985. *Introduction to Artificial Intelligence.* Reading, Mass.: Addison-Wesley.

Churchland, P. 1979. *Scientific Realism and the Plasticity of Mind.* Cambridge: Cambridge University Press.

Churchland, P. 1981. Eliminative materialism and the propositional attitudes. *Journal of Philosophy* 78, no. 2: 67–90.

Churchland, P. 1986. *Neurophilosophy: Towards a Unified Theory of the Mind-Brain.* Cambridge: MIT Press.

Churchland, P. 1989. On the nature of theories: A neurocomputational perspective. In P. M. Churchland, *The Neurocomputational Perspective.* Cambridge: MIT Press.

Churchland, P., and Churchland, P. 1978. Commentary on cognition and consciousness in non-human species. *Behavioural and Brain Sciences* 4: 565–566.

Churchland, P., and Churchland, P. 1981. Functionalism, qualia, and intentionality. In J. Biro and R. Shahan, eds. *Mind, Brain, and Function.* Oklahoma: University of Oklahoma Press, 1982.

Chomsky, N., and Katz, J. 1974. What the linguist is talking about. In N. Block, ed., *Readings in Philosophy of Psychology*, vol. 2, 1980. pp. 223–237. London: Methuen and Co.

Clark, A. 1986. A biological metaphor. *Mind and Language* 1, no. 1: 45–64.
Clark, A. 1987a. From folk-psychology to naive psychology. *Cognitive Science* 11, no. 2: 139–154.
Clark A. 1987b. Connectionism and cognitive science. In J. Hallam and C. Mellish, eds., *Advances in Artificial Intelligence*, pp. 3–15. Chichester: Wiley.
Clark, A. 1987c. The kludge in the machine. *Mind and Language* 2, no. 4: 277–300.
Clark, A. Forthcoming. Thoughts, sentence, and cognitive science. *Philosophical Psychology*.
Cole, M., Hood, L., and McDermott, R. (1978). Ecological niche picking. In U. Neisser, ed., *Memory Observed: Remembering in Natural Contexts*. San Francisco: Freeman, 1982.
Cosmides, L. 1985. Deduction or Darwinian algorithms? An explanation of the "elusive" content effect on the Wason Selection Test. Doctoral thesis. Harvard University.
Davidson, D. 1973. The material mind. In J. Haugeland, ed., *Mind Design*, pp. 339–354. Cambridge: MIT Press, 1981.
Davidson, D. 1984. *Inquiries into Truth and Interpretation* Oxford: Oxford University Press.
Davies, M. 1986. Individualism and supervenience. *Proceedings of the Aristotelian Society*, Suppl. 60: 263–283.
Davies, M. Forthcoming. Modularity: Levels of explanation, neuropsychology and connectionism. Paper presented to the Working Party on Mental Representation. Manchester University, 1987. Forthcoming in *British Journal for the Philosophy of Science*.
Dawkins, R. 1986. *The Blind Watchmaker*. England: Longman.
De Kleer, J., and Brown J. 1985. A qualitative physics based on confluences. In J. Hobbs and R. Moore, eds. *Formal Theories of the Commonsense World*, pp. 109–183. Horwood, N.J.: Ablex.
Dennett, D. 1981. *Brainstorms*. Sussex: Harvester Press.
Dennett, D. 1984a. *Elbow Room: The Varieties of Free Will Worth Wanting*. Oxford: Oxford University Press.
Dennett, D. 1984b. Cognitive wheels: The frame problem of AI. In C. Hookway, ed., *Minds, Machines and Evolution*. Cambridge: Cambridge University Press.
Dennett, D. 1987. *The Intentional Stance*. Cambridge: MIT Press.
Dennett, D. 1988. The evolution of consciousness. Tufts University, Center for Cognitive Studies. Circulating manuscript, CCM 88-1.
Devitt, M., and Sterelny, K. 1987. *Language and Reality: An Introduction to the Philosophy of Language*. Oxford: Blackwell.
Draper, S. 1986. Machine learning and cognitive development. To appear in J. Rutkowska and C. Crook, eds., *The Computer and Human Development: Psychological Issues*.
Dreyfus, H. 1972. *What Computers Can't Do*. New York: Harper and Row.
Dreyfus, H. 1981. From micro-worlds to knowledge representation: AI at an impasse. In J. Haugeland, ed., *Mind Design*, pp. 161–205. Cambridge: MIT Press.
Dreyfus, H., and Dreyfus, S. 1986. *Mind over Machine: The Power of Human Intuition and Expertise in the Era of the Computer*. New York: Free Press, MacMillan.
Durham, T. 1987. Neural brainwaves break new ground. *Computing*, 9 April 1987.
Evans, G. 1982. *The Varieties of Reference* Oxford: Oxford University Press.
Feigenbaum, E. 1977. The art of artificial intelligence: 1. Themes and case studies of knowledge engineering. *Proceedings of the fifth International Joint Conference on Artificial Intelligence* 11: 1014–1029.
Fodor, J. 1968. The appeal to tacit knowledge in psychological explanation. *Journal of Philosophy* 65: 627–640.
Fodor, J. 1980a. Methodological solipsism considered as a research strategy in cognitive psychology. Reprinted in J. Haugeland, ed., *Mind Design*, pp. 307–339. Cambridge: MIT Press, 1981.

Fodor, J. 1980b. Some notes on what linguistics is about. In N. Block, ed., *Readings in Philosophy of Psychology*, vol. 2, pp. 197–207. London: Methuen and Co.
Fodor, J. 1985. Fodor's guide to mental representations: The intelligent Auntie's vade-mecum. *Mind* 94: 77–100.
Fodor, J. 1986. Individualism and supervenience. *Proceedings of the Aristotelian Society*, Suppl. 60: 235–263.
Fodor, J. 1987. *Psychosemantics: The Problem of Meaning in the Philosophy of Mind* Cambridge: MIT Press.
Fodor, J., and Pylyshyn, Z. 1988. Connectionism and cognitive architecture: A critical analysis. *Cognition* 28: 3–71.
Gould, S., and Lewontin R. 1978. The spandrels of San Marco and the Panglossian Paradigm: A critique of the adaptationist programme. Reprinted in E. Sober, ed., *Conceptual Issues in Evolutionary Biology*, Cambridge: MIT Press, 1984.
Hallam, J., and Mellish, C., eds. 1987. *Advances in Artificial Intelligence* Chichester: Wiley and Sons.
Harcourt, A. 1985. All's fair in play and politics. *New Scientist*, no. 1486, December.
Harris, M., and Coltheart, M. 1986. *Language Processing in Children and Adults*. London: Routledge and Kegan Paul.
Haugeland, J. 1981. The nature and plausibility of cognitivism. In J. Haugeland, ed., *Mind Design*, pp. 243–281. Cambridge: MIT Press.
Haugeland, J. 1985. *Artificial Intelligence: The Very Idea* Cambridge: MIT Press.
Hayes, P. 1979. The naive physics manifesto. In D. Michie, ed., *Expert Systems in the Micro-Electronic Age*. Edinburgh: Edinburgh University Press.
Hayes, P. 1985a. The second naive physics manifesto. In J. Hobbs and R. Moore, eds., *Formal Theories of the Commonsense World*, pp. 1–36. Norwood, N.J.: Ablex.
Hayes, P. 1985b. Naive physics I: Ontology for liquids. In J. Hobbs and R. Moore, eds., *Formal Theories of the Commonsense World* Norwood, N.J.: Ablex.
Hebb, D. 1949. *The Organization of Behavior*. New York: Wiley and Sons.
Hinton, G. 1984. Parallel computations for controlling an arm. *Journal of Motor Behavior* 16: 171–194.
Hinton, G., and Anderson, J. 1981, eds. *Parallel Models of Associative Memory*. Hillsdale, N.J.: Erlbaum.
Hobbs, J. 1985. Introduction to J. R. Hobbs and R. Moore, eds., *Formal Theories of the Commonsense World*, pp. xi–xxii. Norwood, N.J.: Ablex.
Hobbs, J., and Moore, R. 1985. eds. *Formal Theories of the Commonsense World*. Norwood, N.J.: Ablex.
Hodges, A. 1983. *Alan Turing: The Enigma*. New York: Simon and Schuster.
Hofstadter, D. 1985. Waking up from the Boolean dream, or, Subcognition as computation. In his *Metamagical Themas: Questing for the Essence of Mind and Pattern*, pp. 631–665. Harmondsworth: Penguin.
Hornsby, J. 1986. Physicalist thinking and behaviour. In P. Pettit and J. McDowell, eds., *Subject, Thought, and Context*. Oxford: Oxford University Press.
Hull, D. 1984. Historical entities and historical narratives. In C. Hookway, ed., *Minds, Machines and Evolution*. Cambridge: Cambridge University Press.
Humphreys, N. 1983. Nature's psychologists. *Consciousness Regained*. New York: Oxford University Press.
Israel, D. 1985. A short companion to the naive physics manifesto. In J. Hobbs and R. Moore, eds., *Formal Theories of the Commonsense World*, pp. 427–447. Norwood, N.J.: Ablex.
Jacob, F. 1977. Evolution and tinkering. *Science* 196, no. 4295: pp. 1161–1166.
Jackson, F., and Pettit, P. 1988. Functionalism and Broad Content. *Mind* 97, no. 387: 381–400.

Kahneman, D., Slovic, P., and Tversky, A., eds. 1982. *Judgement under Uncertainty: Heuristics and Biases.* Cambridge: Cambridge University Press.
Karmiloff-Smith, A. 1984. Children's problem solving. In M. E. Lamb, A. L. Brown, and B. Rogoff, eds. *Advances in Developmental Psychology*, vol. 3, pp. 39–90. Hillsdale, N.J.: Erlbaum.
Karmiloff-Smith, A. 1985. Language and cognitive processes from a developmental perspective. *Language and Cognitive Processes* 1, no. 1: p. 61–85.
Karmiloff-Smith, A. 1986. From metaprocesses to conscious access: Evidence from children's metalinguistic and repair data. *Cognition* 23: 95–147.
Karmiloff-Smith, A. 1987. Beyond modularity: A developmental perspective on human consciousness. Draft manuscript of a talk given at the annual meeting of the British Psychological Society, Sussex, April.
Katz, J. 1964. Mentalism in linguistics. *Language* 40: 124–137.
Krellenstein, M. 1987. A reply to parallel computation and the mind-body problem. *Cognitive Sciences* 11: 155–157.
Knuth, D. 1973. *Sorting and Searching.* Reading, Mass.: Addison-Wesley.
Köhler, W. 1929. *Gestalt Psychology.* New York: Liveright.
Kuczaj, S. A. 1977. The acquisition of regular and irregular past tense forms. *Journal of Verbal Learning and Verbal Behaviour* 16: 589–600.
Lakatos, I. 1974. Falsification and the methodology of scientific research programmes. In I. Laktos and A. Musgrave, eds., *Criticism and the Growth of Knowledge.* Cambridge: Cambridge University Press.
Langley, P. 1979. Rediscovering physics with BACON 3. *Proceedings of the Sixth International Joint Conference on Artificial Intelligence* 1: 505–508.
Langley, P., Simon, H., Bradshaw, G., and Zytkow, J. 1987. *Scientific Discovery: Computational Explorations of the Creative Process.* Cambridge: MIT Press.
Lenat, D. 1977. The ubiquity of discovery. *Proceedings of the fifth International Joint Conference on Artificial Intelligence* 2: 1093–1105.
Lenat, D. 1983a. Theory formation by heuristic search. *Artificial Intelligence* 21: 31–59.
Lenat, D. 1983b. EURISKO: A program that learns new heuristics and domain concepts. *Artificial Intelligence* 21: 61–98.
Levi-Strauss, C. 1962. *The Savage Mind.* London: Weidenfeld and Nicolson.
Lieberman, P. 1984. *The Biology and Evolution of Language.* Cambridge: Harvard University Press.
Lycan, W. 1981. Form, function, and feel. *Journal of Philosophy* 78, no. 1: 24–50.
Maloney, J. 1987. The right stuff. *Synthese* 70: 349–372.
McClelland, J. 1981. Retrieving general and specific knowledge from stored knowledge of specifics. *Proceedings of the Third Annual Conference of The Cognitive Science Society* (Berkeley) 170–172.
McClelland, J. 1986. The programmable blackboard model of reading. In J. McClelland, D. Rumelhart, and the PDP Research Group, *Parallel Distributed Processing: Explorations in the Microstructure of Cognition*, vol. 2 pp. 122–169. Cambridge: MIT Press.
McClleland, J., and Kawamoto, A. 1986. Mechanisms of sentence processing: Assigning roles to constituents of sentences. In J. McClelland, D. Rumelhart, and the PDP Research Group, *Parallel Distributed Processing: Explorations in the Microstructure of Cognition*, vol. 2, pp. 216–271. Cambridge: MIT Press.
McClelland, J., and Rumelhart, D. 1985a. Distributed memory and the representation of general and specific information. *Journal of Experimental Psychology: General* 114, no. 2: 159–188.
McClelland, J., and Rumelhart, D. 1985b. Levels indeed! A response to Broadbent. *Journal of Experimental Psychology: General* 114, no. 2: 193–197.

McClelland, J., and Rumelhart, D. 1986. Amnesia and distributed memory. In J. McClelland, D. Rumelhart, and the PDP Research Group, *Parallel Distributed Processing: Explorations in the Microstructure of Cognition*, vol. 2, p. 503–529. Cambridge: MIT Press.

McClelland, J., Rumelhart, D., and Hinton, G., 1986. The appeal of PDP. In Rumelhart, McClelland, and the PDP Research Group. *Parallel Distributed Processing: Explorations in the Microstructure of Cognition*, vol. 1, pp. 3–44. Cambridge: MIT Press.

McClelland, J., and Rumelhart, D., and the PDP Research Group, *Parallel Distributed Processing: Explorations in the Microstructure of Cognition*, vol. 2 Cambridge: MIT Press.

McCulloch, G. 1986. Scientism, mind, and meaning. In P. Pettit and J. McDowell, eds., *Subject, Thought, and Context*. Oxford: Oxford University Press. 1986.

McCulloch, W., and Pitts, W. 1943. A logical calculus of the ideas immanent in nervous activity. *Bulletin of Mathematical Biophysics* 5: 115–133.

McDermott, D. 1976. Artificial intelligence meets natural stupidity. In J. Haugeland, ed., *Mind Design* Cambridge: MIT Press, 1981.

McGinn, C. 1982. The structure of content. In A. Woodfield, ed., *Thought and Object*, pp. 207–259. Oxford: Oxford University Press.

Marr, D. 1977. Artificial intelligence: A personal view. In J. Haugeland, ed., *Mind Design*, p. 129–142. Cambridge: MIT Press, 1981.

Marr, D. 1982. *Vision*. New York: W. H. Freeman and Co.

Marr, D., and Poggio, T. 1976. Cooperative computation of stereo disparity. *Science* 194: 283–287.

Michaels, C., and Carello, C. 1981. *Direct Perception*. Englewood Cliffs, N.J.: Prentice-Hall.

Michie, D., and Johnston, R. 1984. *The Creative Computer*. Harmondsworth: Penguin.

Millikan, R. 1986. Thoughts without laws, cognitive science with content. *Philosophical Review*. 95: 47–80.

Minsky, M. 1974. A framework for representing knowledge. MIT lab memo 306. Cambridge, Mass. Excerpts in J. Haugeland, ed., *Mind Design* (Cambridge: MIT Press, 1981).

Minsky, M., 1980. K-lines: A theory of memory. *Cognitive Science* 4: 117–133.

Minsky, M., and Papert, S. 1969. *Perceptrons*. Cambridge: MIT Press.

Newell, A. 1980. Physical symbol systems. *Cognitive Science* 4: 135–183.

Newell, A., and Simon, H. 1976. Computer science as empirical inquiry. In J. Haugeland, ed., *Mind Design*. Cambridge: MIT Press.

Norman, D. 1986. Reflections on cognition and parallel distributed processing. In J. McClelland, D. Rumelhart, and the PDP Research Group, *Parallel Distributed Processing: Explorations in the Microstructure of Cognition*, vol. 2, pp. 110–146. Cambridge: MIT Press.

Pettit, P., and McDowell, J., eds., 1986. *Subject, Thought and Context*. Oxford: Oxford University Press.

Pinker, S. 1984. *Language Learnability and Language Development*. Cambridge: Harvard University Press.

Pinker, S., and Prince. A. 1988. On language and connectionism: Analysis of a parallel distributed processing model of language acquisition. *Cognition* 28: 73–193.

Poggio, T., and Koch, C. 1987. Synapses that compute motion. *Scientific American*, May, pp. 42–48.

Premack, D., and Woodruff, G. 1978. Does the chimpanzee have a theory of mind? *Behavioural and Brain Science* 4: 515–526.

Putnam, H. 1960. Minds and machines. In S. Hook, ed., *Dimensions of Mind*. New York: New York University Press.

Putnam, H. 1967. Psychological Predicates. In W. Capitan and D. Merill, eds., *Art, Mind, and Religion*, pp. 37–48. University of Pittsburgh Press.

Putnam, H. 1975a. The meaning of "meaning." In H. Putnam, *Mind, Language, and Reality,* pp. 215–271. Cambridge: Cambridge University Press.
Putnam, H. 1975b. Philosophy and our mental life. In H. Putnam, *Mind, Language and Reality,* pp. 291–303. Cambridge: Cambridge University Press.
Putnam, H. 1981. Reductionism and the nature of psychology. In J. Haugeland, ed., *Mind Design,* pp. 205–219. Cambridge: MIT Press.
Pylyshyn, Z. 1986. *Computation and Cognition.* Cambridge: MIT Press.
Ridley, M. 1985. *The Problems of Evolution.* Oxford: Oxford University Press.
Ritchie, G., and Hanna, F. 1984. AM: A case study in AI methodology. *Artificial Intelligence* 23: 249–268.
Robbins, A. Unpublished. Representing type and category in PDP. Draft doctoral dissertation. University of Sussex.
Rosenblatt, F. 1962. *Principles of Neurodynamics.* New York: Spartan Books.
Rumelhart, Hinton, G., and Williams, R. Learning internal representations by error propagation. In Rumelhart, McClelland, and the PDP Research Group, *Parallel Distributed Processing: Explorations in the Microstructure of Cognition,* vol. 1, pp. 318–362. Cambridge: MIT Press.
Rumelhart, D., and McClelland, J. 1986. On learning the past tenses of English verbs. In J. McClelland, D. Rumelhart, and the PDP Research Group, *Parallel Distributed Processing: Explorations in the Microstructure of Cognition,* vol. 2, pp. 216–271. Cambridge: MIT Press.
Rumelhart, D., and McClelland, J. 1986. PDP models and general issues in cognitive science. In D. Rumelhart, J. McClelland, and the PDP Research Group, *Parallel Distributed Processing: Explorations in the Microstructure of Cognition,* vol. 1, pp. 110–146. Cambridge: MIT Press.
Rumelhart, D., Mclelland, J., and the PDP Research Group, 1986. *Parallel Distributed Processing: Explorations in the Microstructure of Cognition,* vol. 1, Cambridge: MIT Press.
Rumelhart, D., and Norman, D. 1982. Simulating a skilled typist: A study in skilled motor performance. *Cognitive Science* 6: 1–36.
Rumelhart, D., Smolensky, P., McClelland, J., and Hinton, G. 1986. Schemata and sequential thought processes in PDP models. In J. McClelland, D. Rumelhart, and the PDP Research Group, *Parallel Distributed Processing: Explorations in the Microstructure of Cognition,* vol. 2, pp. 7–58. Cambridge: MIT Press.
Rutkowska, J. 1984. Explaining infant perception: Insights from artificial intelligence. Cognitive studies research paper 005. University of Sussex.
Rutkowska, J. 1986. Developmental psychology's contribution to cognitive science. In K.S. Gill, ed. *Artificial Intelligence for Society,* pp. 79–97. Chichester, Sussex: John Wiley.
Ryle, G. 1949. *The Concept of Mind.* London: Hutchinson.
Sacks, O. 1986. *The Man Who Mistook His Wife for a Hat.* London: Picador.
Schank, R., and Abelson, R. 1977. *Scripts, Plans, Goals, and Understanding.* Hillsdale, N.J.: Lawrence Erlbaum Associates.
Schilcher, C., and Tennant, N. 1984. *Philosophy, Evolution, and Human Nature.* London: Routledge and Kegan Paul.
Schreter, Z., and Maurer, R. 1986. Sensorimotor spatial learning in connectionist artificial organisms. Research abstract FPSE. University of Geneva.
Searle, J. 1969. *Speech Acts: An Essay in the Philosophy of Language.* Cambridge: Cambridge University Press.
Searle, J. 1980. Minds, brains, and programs. Reprinted in J. Haugeland, ed., *Mind Design,* pp. 282–307. Cambridge: MIT Press, 1981.
Searle, J. 1983. *Intentionality.* Cambridge: Cambridge University Press.

Searle, J. 1984. Intentionality and its place in nature. *Synthese* 61: 3–16.

Sejnowski, T., and Rosenberg, C. 1986. NETtalk: A parallel network that learns to read aloud. John Hopkins University Technical Report JHU/EEC-86/01.

Shortlife, E. 1976. *Computer Based Medical Consultations: MYCIN* New York: Elsevier.

Simon, H. 1962. The architecture of complexity. Reprinted in H. Simon, ed., *The Sciences of the Artificial.* Cambridge: Cambridge University Press, 1969.

Simon, H. 1979. Artificial intelligence research strategies in the light of AI models of scientific discovery. *Proceedings of the Sixth International Joint Conference on Artificial Intelligence* 2: 1086–1094.

Simon, H. 1980. Cognitive science: The newest science of the artificial. *Cognitive Science* 4, no. 2: 33–46.

Simon, H. 1987. A psychological theory of scientific discovery. Paper presented at the annual conference of the British Psychological Society. University of Sussex.

Sloman, A. 1984. The structure of the space of possible minds. In S. Torrance, ed., *The Mind and The Machine.* Sussex: Ellis Horwood.

Smart, J. 1959. Sensations and brain processes. *Philosophical Review* 68: 141–156.

Smith, M. 1984. The evolution of animal intelligence. In C. Hookway, ed., *Minds, Machines and Evolution.* Cambridge: Cambridge University Press.

Smolensky, P. 1986. Information processing in dynamical systems: Foundations of harmony theory. In D. Rumelhart, J. McClelland, and the PDP Research Group. *Parallel Distributed Processing: Explorations in the Microstructure of Cognition,* vol. 1, pp. 194–281. Cambridge: MIT Press.

Smolensky, P. 1987. Connectionist AI, and the brain. *Artificial Intelligence Review* 1: 95–109.

Smolensky, P. 1988 On the proper treatment of connectionism. *Behavioural and Brain Sciences* 11: 1–74.

Sterelny, K. 1985. Review of Stich *From Folk Psychology to Cognitive Science. Australasian Journal of Philosophy* 63, no. 4: 510–520.

Stich, S. 1971. What every speaker knows. *Philosophical Review* 80: 476–496.

Stich, S. 1972. Grammar, psychology, and indeterminacy. Reprinted in N. Block, ed., *Readings in Philosophy of Psychology,* vol. 2, pp. 208–222. London: Methuen and Co., 1980.

Stich, S. 1983. *From Folk Psychology to Cognitive Science.* Cambridge: MIT Press.

Tannenbaum, A. 1976. *Structured Computer Organization.* Englewood Cliffs, N.J.: Prentice-Hall.

Tennant, N. 1984a. Intentionality, syntactic structure, and the evolution of language. In C. Hookway, ed., *Minds, Machines, and Evolution.* Cambridge: Cambridge University Press.

Tennant, N., and Schilcher, C. 1984. *Philosophy, Evolution, and Human Nature.* London: Routledge and Kegan Paul.

Tennant, N. 1987. Philosophy and biology: Mutual enrichment or one-sided encroachment. *La nuova critica* 1–2: 39–55.

Thagard, P. 1986. Parallel computation and the mind-body problem. *Cognitive Science* 10: 301–318.

Torrance, S. 1984. Philosophy and AI: Some issues. Introduction to S. Torrance, ed., *The Mind and the Machine,* pp. 11–28. Sussex: Ellis Horwood.

Turing, A. 1937. On computable numbers with an application to the Entscheidungs problem. *Proceedings of the London Mathematical Society* 42: 230–265.

Turing, A. 1950. Computing machinery and intelligence. *Mind* 59: 433–460.

Van Fraasen, B. 1980. *The Scientific Image.* Oxford: Oxford University Press.

Vogel, S. 1981. Behaviour and the physical world of an animal. In P. Bateson and P. Klopfer, eds., *Perspectives in Ethology,* vol. 4. New York: Plenum Press.

Walker, S. 1983. *Animal Thought.* London: Routledge and Kegan Paul.

Warrington, C., and McCarthy, R., 1987. Categories of knowledge: Further fractionations and an attempted integration. *Brain* 110: 1273–1296.
Winograd, T. 1972. Understanding natural language. *Cognitive Psychology* 1: 1–191.
Winston, P. 1975. *The Psychology of Computer Vision*. New York: McGraw Hill.
Wittgenstein, L. 1969. *On Certainty*. Oxford: Blackwell.
Woodfield, A., ed. 1982. *Thought and object*. Oxford: Oxford University Press.

Index